"Brottman writes amusingly and often movingly of the relationships between dozens of writers and artists and their canine friends, along the way exploring her own devotion to Grisby, a charismatic if at times bumptious French bulldog. . . . Brottman's research is deep and her storytelling compelling."

—*Boston Globe*

"A history of smart, important women and men and their dogs. . . . Incredibly touching. . . . Any reader who takes the time to put aside dog-book expectations will love it for the way it can both move you and challenge you. . . . A life-affirming book for dog owners."

—*Baltimore City Paper*

"If *The Great Grisby* were a meal, it would satisfy (at last) even the hungriest of dogs and the most discerning of passionate dog-lovers. No kibble, a feast. I devoured it in one savory bite."

—Abigail Thomas, author of *A Three Dog Life*

"I have read thousands of books in my eighty-one years, and this is the only one that has made me happy. . . . Learned, spritely."

—Jonathan Mirsky, *The Spectator* (London),
a book of the year

"Amazing. . . . If you love animals and are an aficionado of dogs this is a wonderful look at the trials and tribulations of those in the past as they come to terms with the differing personalities of the pets they call their own."   —*Seattle Post-Intelligencer*

"Part history book and part memoir, *The Great Grisby* is a fascinating exploration of how dogs have changed people and the world in myriad ways."   —*Shelf Awareness*

"In her paeans to her pet, Brottman evokes the joys of dog ownership. . . . Avid dog-lovers will relish the digressions into literature and history, as well as the assurance that the love between dog and human can be as deep as any other kind of love."

—*Publishers Weekly*

"The greatest pleasure of this wonderful book is reading about the author's love for Grisby. . . . Her descriptions of the two of them spending the day quietly at home is mesmerizing in its ability to remind us of the simple pleasure of hanging together, dog and person."

—Jeffrey Moussaieff Masson, author of *Beasts* and *Dogs Never Lie About Love*

"A quirky, funny, scholarly romp of a book . . . chock-full of wonderful trivia and little-known tidbits. . . . There was not one chapter that didn't leave me smiling, laughing, or gasping in amazement."

—Booksforanimallovers.com

"Unusual and remarkable. . . . [There are] more questions than answers, and yet somehow a sense of understanding, profound understanding, emerges from these queries. It seems to take a dog to bring us to this point of wordless wisdom. . . . Highly recommended."

—VegetarianFriends.net

"More than just a collection of fun dog stories, Brottman's book illuminates the symbiotic relationship between people and canines, offering insights into the human condition through the lens of our four-footed friends."

—*Booklist*

The Great Grisby

THE
GREAT
GRISBY

TWO THOUSAND YEARS OF
LITERARY, ROYAL, PHILOSOPHICAL,
AND ARTISTIC DOG LOVERS AND
THEIR EXCEPTIONAL ANIMALS

MIKITA BROTTMAN

Illustrated by Davina "Psamophis" Falcão

HARPER ● PERENNIAL

NEW YORK • LONDON • TORONTO • SYDNEY • NEW DELHI • AUCKLAND

HARPER ● PERENNIAL

A hardcover edition of this book was published in 2014 by HarperCollins Publishers.

HarperCollins books may be purchased for educational, business, or sales promotional use. For information please e-mail the Special Markets Department at SPsales@harpercollins.com.

FIRST HARPER PERENNIAL EDITION PUBLISHED 2015.

Illustrations by Davina "Psamophis" Falcão

Designed by Fritz Metsch

THE LIBRARY OF CONGRESS HAS CATALOGUED THE HARDCOVER EDITION AS FOLLOWS:

BROTTMAN, MIKITA.
THE GREAT GRISBY : TWO THOUSAND YEARS OF LITERARY, ROYAL, PHILOSOPHICAL, AND ARTISTIC DOG LOVERS AND THEIR EXCEPTIONAL ANIMALS / BY MIKITA BROTTMAN ; ILLUSTRATED BY DAVINA "PSAMOPHIS" FALCÃO.—FIRST EDITION.
PAGES CM
ISBN 978-0-06-230461-2
1. DOGS. 2. ANIMAL BEHAVIOR. I. TITLE.
SF426.2.B75 2014
636.7—DC   2013048448

ISBN 978-0-06-230462-9 (pbk.)

15 16 17 18 19   OV/RRD   10 9 8 7 6 5 4 3 2 1

A dog starv'd at his master's gate
Predicts the ruin of the state.

—WILLIAM BLAKE,
"Auguries of Innocence" (1803)

Contents

	Introduction	1
1.	*Atma*	5
2.	*Bull's-eye*	13
3.	*Caesar III*	21
4.	*Douchka*	29
5.	*Eos*	39
6.	*Flush*	47
7.	*Giallo*	57
8.	*Hachikō*	65
9.	*Issa*	73
10.	*Jip*	81
11.	*Kashtanka*	89
12.	*Lump*	97
13.	*Mathe*	107
14.	*Nero*	115
15.	*Ortipo*	123
16.	*Peritas*	135
17.	*Quinine*	143
18.	*Robber*	153
19.	*Shock*	165
20.	*Tulip*	175
21.	*Ulisses*	185

Contents

22. *Venom* 193

23. *Wessex* 201

24. *Xolotl* 209

25. *Yofi* 217

26. *Zémire* 225

Postscript 233

Notes 235

Bibliography 259

Introduction

"UNABLE TO LOVE each other," writes the British author J. R. Ackerley, "the English turn naturally to dogs." I acquired my first dog when I was close to forty, and my eight-year love affair with this willful and charismatic animal has led me to wonder whether it's true, as Ackerley suggests, that there's something repressed and neurotic about those whose deepest feelings are for their dogs. Thinking about this question has led me not to an answer but to further questions. Is my relationship with Grisby nourishing or dysfunctional, commonplace or unique? Do we choose and train dogs in our own image? Why are some people drawn to poodles, some to bulldogs, and others to dachshunds? Can devotion to a dog become pathological? Why is a woman's love for her lapdogs considered embarrassingly sentimental when men bond so proudly with their well-built hounds? Married women admit they sleep with their dogs, and married men deny it; someone's not telling the truth, but who's lying, and why? What drives some people to wash their hands obsessively after any canine contact while others are happy to share flatware with Fido? And why is "Fido" still used as the generic dog's name when it's been out of fashion for almost a hundred years?

Each of this book's twenty-six chapters is devoted to a particular human-canine bond. Some of these couplings are drawn from literature, where dogs are generally symbolic, often standing as their owners' avatars, sharing similar characteristics or drawing attention to vital clues that the human characters have overlooked. Other pairings are drawn from history, art, folklore, and philosophy, and cover a broad span of history (320 BC to the 1970s) and geography (Rome, Russia, Japan, Germany, Mexico, Malta, Greece, the United States), but particular attention is paid to dog-and-owner pairs from Victorian Britain. According to authorities on the subject, late-nineteenth-century England saw the origins of modern dog breeding and pet keeping, which led to an increase in the depiction of dogs in art and literature, as well as their increase in everyday life, among all classes and age groups.

Far apart as these human-dog stories may be in time and place, their themes are remarkably consistent. Exceptional dogs, it turns out, often have traits in common, and the most familiar of these is miraculous loyalty. History and folklore are full of dogs that won't leave their owners' dead or injured bodies; dogs that spend every night at their masters' graves; dogs that drown themselves in grief, conceal themselves under their mistresses' skirts as they're led to the scaffold, or travel thousands of miles to make their way home. The fact that such tales have become folklore does not mean they are not also true. Dogs are remarkably faithful creatures. Upon further investigation, however, these miraculously loyal dogs often turn out to be rather less miraculous than their stories suggest, though no less interesting for that.

Many of the dogs described in this book will be unfamiliar to the reader, and I'm especially interested in these lesser-known dogs. A lot has been said and written already about iconic, mediagenic dogs like Lassie, Old Yeller, and Rin Tin Tin. In *The Great Grisby*, I draw attention to dogs that inhabit the margins or lurk on the periphery, dogs that have been overlooked. As is so often the case, those who are allowed behind the scenes or on the sidelines (children, servants, janitors, busboys) often get to see and experience things that are normally kept from public view. Partly because they can't speak but mainly because they don't judge, dogs have unfettered access to the backstage of life. Imagine what Prince Albert's dog Eos could have told us about Queen Victoria, or what Freud's dog Yofi might have learned from his master's patients. A dog in the room is a silent observer, a witness to the human drama: it sees all, smells all, and says nothing.

All the dogs described in this book are, like Grisby, exceptional. This obviously raises the question of what makes an exceptional dog. The answer is simple. What makes a dog exceptional is its owner. In other words, any dog can be exceptional if it's loved enough. We see our dogs through human eyes; this is the transformative power of projection. In order to understand this process more fully, I don my psychoanalytic hat and, taking a cue from Freud (another late-life dog lover), I put the human-canine relationship on the couch (never mind the dog hair). The way we think about our dogs is infinitely revealing. Rich insights can be gained from observing how people name their dogs, create personalities for them, address them, speak on their behalf, even from the way they pick up after them. For some, a dog is an alter ego; for others, a substitute for a child; other people use their dogs to keep the world at bay, to heal wounds inflicted in infancy, or to recapture their playful, preverbal selves.

The Great Grisby is structured like a leisurely stroll in the park. We begin with Atma, the name given by the misanthropic philosopher Arthur Schopenhauer to his succession of standard poodles, and continue alphabetically until we arrive at Zémire, the adored pet of the French poet and intellectual Madame Antoinette Deshoulières. However, the path is not always direct. In this book, as on all our walks, Grisby sometimes leads us on sidetracks, following scents, sniffing out clues and connections, retracing our steps, taking us into the realms of folklore, semiotics, philosophy, and zoology. Sometimes he seems to be leading us astray, but as long as we're together, we'll never be lost. Everywhere, every day, he shows me how dog is the mirror of man.

A T M A

THE FAMOUSLY MISANTHROPIC German philosopher Arthur Schopenhauer spent twenty-seven years of his life living alone, averse to human company, but like other notorious malcontents, he was deeply attached to his dogs. Throughout his life, from his student days at Göttingen until his death at Frankfurt am Main, Schopenhauer owned a succession of standard poodles—a famously loyal, active, and intelligent breed. "To anyone who needs lively entertainment for the purpose of banishing the dreariness of solitude," he wrote in 1851, "I recommend a dog, in whose moral and intellectual qualities he will almost always experience delight and satisfaction."

Though he remained loyal to the standard poodle, the philosopher's companions varied in color. The dog he owned in the 1840s was white, and the one he owned at his death—and for which he provided generously in his will—was brown. According to the few guests who visited his home, Schopenhauer was deeply attentive to these animals; though his daily routine was rigid, he always made sure his poodles got regular constitutionals. He even concerned himself with their daily amusements. One colleague recalled being in the middle of an earnest conversation with the philosopher at his home when they were interrupted by the music of a regimental band passing the window, at which point Schopenhauer got up and moved his poodle's seat closer, to give him a better view of the procession.

The philosopher was ahead of his time in his concern for animal suffering. "When I see how man misuses the dog, his best friend; how he ties up this intelligent animal with a chain," he wrote, "I feel the deepest sympathy with the brute and burning indignation against its master." Yet curiously, while he respected his dogs as individuals, Schopenhauer gave every one of them the same name: Atma (though his last dog—the brown one— generally went by the nickname "Butz"). *Atma* is the Hindu word for the universal soul (or, as Schopenhauer interpreted it, the impersonal, primordial, eternally renewed force of nature). Historians of philosophy have suggested this naming habit may be connected to Schopenhauer's theory of individuality, and his notion that a particular type of animal expresses the Platonic ideal of its kind.

It may, on the other hand, have been something more familiar: an attempt to forestall the pain of loss. This is why the authors Gertrude Stein and Alice B. Toklas gave their new dog—also, coincidentally, a standard poodle—the same name as the dog they'd just lost: Basket. The first Basket was an elegant creature acquired by Toklas at a Parisian dog show, and so named because she immediately pictured him carrying a basket of flowers in his mouth (a skill he never fully mastered). The poodle was treated like a young prince, bathed daily in sulfur water for the benefit of his sensitive skin. Stein let him sit in her lap when she wrote ("on Mount Gertrude," said Alice), and she claimed the rhythm of the dog's breathing taught her the essential difference between sentences and paragraphs. When Basket died in 1937, the bereaved women, on the advice of a friend, acquired a similar-looking dog, and gave him the same name.

When the second Basket arrived in 1938, the couple were living in France. Despite widespread rationing, Basket II didn't go hungry. The Nazis' theory of racial purity extended even to pets; as long as a dog had a documented pedigree, it received a food allowance. Basket II lived fourteen years, six past the death of Stein herself, and when he died, Alice B. Toklas was left alone. "His going has stunned me," she wrote to her friend Carl Van Vechten. "For some time I have realized how much I depended upon him and so it is the beginning of living for the rest of my days without anyone who is dependent on me for anything." She was too old, she reasoned, to acquire a Basket III.

It's natural that a bereaved pet owner should want to stave off the pain of loss by acquiring a second dog that resembles the original, and while I can understand the impulse to name a new dog after a beloved old one, the scheme has a major flaw. In my experience, whatever their breed, dogs are unique and as individual as humans, and you can't make a gentle dog tough just by calling him Butch. Ideally, we should wait until we've gotten a sense of a dog's personality before picking out a name, but puppy owners, like would-be parents, usually have a name in mind long before they lay eyes on their new arrival.

Grisby is the first and only dog I've ever owned, and I had his name picked out before he was even born. One evening, my partner, David, and I watched a French movie from 1954 called *Touchez Pas au Grisbi* (translation: *Don't Touch the Loot*). The film is about a band of world-weary French gangsters who sit around in a bar planning a heist and mumbling about *le grisbi* (old-fashioned French criminal slang whose equivalent is some-thing like "loot" or "booty"). As I recall, we both thought the word would be an appropriate name for the dog we were plan-ning to acquire, not only because it is French (though we've An-glicized the spelling) but also because it's a tough, macho way of saying "treasure," perfect for a little French bulldog.

The first time I ever laid eyes on one of these creatures, I was walking through Greenwich Village, which of all areas in the United States contains perhaps the greatest concentration of the breed. I was immediately intrigued and enchanted by this odd little animal, with its flat snout and prominent ears. I wanted one so badly it hurt, though it would be another six years before I could fit a dog into my life. Then, when the time was right, I logged on to PuppyFind.com, and there was "Oliver"—a tiny dog with enormous ears and an endearingly inquisitive expression. He was, I thought, the sweetest-looking puppy I'd ever seen, and he'd be weaned by the middle of August, which was exactly when we'd be ready for him. I e-mailed David the photo, though it was a symbolic gesture only—I already knew he was *the one*. By the end of the day, our deposit to the breeders had been paid, and "Oliver"—all ears—was all ours.

As of the time of writing, Grisby is almost eight years old, and he tips the scales at thirty-two pounds. His color is officially designated "fawn piebald," which means he has very pretty markings of light brown and white, about half of each. His fur is short and soft, and his large, expressive ears are light brown on the back, dark pink inside, and can seem almost translucent in the sunlight. He has a stocky, muscular body, no snout to speak of, and no tail. His eyes are deep brown, and one of them has a strabismus, meaning that it looks slightly to the left. His nose and mouth area are black, and like most bulldogs, he has a pro-nounced underbite. His mouth is wide and, when he's trotting along with his pink tongue hanging out, forms a permanent smile. His face is joyful, his eyes bright, his expression either playful or craven.

I could never have imagined that Grisby's name, chosen almost at random, would come to be so full of meaning for me. "You will likely call your dog's name over 50,000 times," advises the author of *How to Raise and Train a French Bulldog*. "Pick a name you like!" In his book *Bashan and I*, the German writer Thomas Mann writes that of all the pleasures he shares with his dog, none is so great for him as addressing the creature over and over again by his name. "Bashan" is the only word that Mann's devoted and playful setter seems to understand, and his master loves driving him into crazy fits of ecstasy reminding him that not only is his *name* Bashan but he *is* Bashan, a truth the dog never seems to tire of. Grisby feels the same way; he seems to love his special name as much as I love to say it. Of course, now that we've been together for eight years, I can't separate the name from the animal it signifies, and I'm irritated when people who've known him for years still haven't grasped it, calling him Grigsby, Bigsby, Gribley, or Grimsby.

We name our dogs the way we name our children; we name the child we imagine having—the child we want—rather than the child we get. Bearing this in mind, there's a lot to learn about people from the names they give their dogs. Some prefer a name they've heard before; others pick something they consider unique, as I did. As with baby names, fashions in dog names go in cycles. In ancient Roman households, it was trendy to give Greek names to your hounds (and your slaves). The most popular Roman dog names were descriptive: Ferox ("Savage"), Melampus ("Blackfoot"), Patricius ("Noble"), and Skylax ("Puppy"). Greek dogs were rarely saddled with the polysyllabic names of their owners (Agamemnon, Olympiodorus). Xenophon, a Greek historian who wrote about hounds in the fourth century BC, maintained that the best names are short, consisting of no more than one or two syllables, so the dogs may be easily called. Popular names were those that expressed speed, courage, and strength, such as Aura ("Breeze"), Horme ("Eager"), Korax ("Raven"), and Labros ("Fierce").

In the United States, until around fifty years ago, dogs were generally working animals rather than household pets, and their names reflected their tasks and talents: Hunter, Skipper, Pilot, Sailor, Shep. Simple, one-syllable names like Buck, Lad, Jack, and Pal are still popular for working dogs; they're easy for the animals to learn, and the owners to yell. Grandiose names like Caesar, Nero, and Napoleon have always been in fashion among purebred pets (and people), and descriptive names—Patch, Jet, Domino, Ginger—are still sometimes heard, though not as much as they once were. Traditional dog names like Rover and Fido are also out of fashion; these days, dogs seldom rove, and few of us speak Latin.

Today, at least in Europe and the United States, very few dogs are kept as working animals. Most pooches live in the home and sleep in their owners' beds; their only task is to provide affection and attention, and they succeed like never before. According to a recent survey, 15 percent of British dog owners consider their pet more important than their cousin, and 6 percent confessed they even preferred their pet to their own partner. Sixteen percent listed their dogs as household members in the 2011 British census, some listing a dog as their "son" on the official form. The deeper the bond we form with our dogs, it seems, the more we make them over in our own image; in keeping with their role as full family members, dogs are now commonly given human names. Today, for the first time in history, the same names turn up in top-ten lists for both babies and dogs: Chloe, Bella, and Sophie for girls; Charlie, Jack, and Max for boys. The same trend is common in Europe, except in the more strictly Catholic countries, where it's considered sacrilegious to call "soulless animals" by human names.

One fashion that hasn't changed over time is the tendency for macho guys to give their tough dogs fighting names. Popular names for male pit bull terriers include Tyson, Diesel, and Tank. Other common names include Chaos, Sherman, and Panzer. In Emily Brontë's novel *Wuthering Heights*, published in 1847, Heathcliff's bulldogs, which he warns are "not kept for a pet," are named Skulker and Throttler. It's not surprising that he mistreats them, nor that he almost kills a spaniel, nor that a child raised in his home is seen "hanging a litter of puppies from a chair-back in the doorway." Puppies that recover from such misfortunes are invariably named Lucky or Chance. Rocky is the most popular name for dogs that bite, according to San Francisco Health Department records, closely followed by Mugsy, Max, and Zeke.

We fall in love with individual dogs, but it's difficult to separate the dog from the breed, and it's not unusual, with dogs as with lovers, that we should fall repeatedly for the same kind. The standard poodle has always been popular with literary and philosophical types. In addition to Schopenhauer and Gertrude Stein, poodle lovers include Victor Hugo, Lillian Hellman, George Sand, Norman Mailer—who once got into a street brawl with a man who called his poodle "a queer"—and John Steinbeck, whose standard poodle Charley was his close companion for many years, and costar of his 1962 book *Travels with Charley*. However, outranking even the poodle among literary and artistic types is the dachshund (see LUMP and QUININE), breed of choice for Henry James, Matthew Arnold, Dorothy Parker, G. K. Chesterton, Anton Chekhov, Vladimir Nabokov, Pablo Picasso, and David Hockney, among others. Dachshunds are said to be complex, vulnerable, and fussy, and they're often described as having an "artistic temperament." Like dog, like master.

In light of my feelings for Grisby, I find it hard to imagine owning any breed other than the bulldog. As the only dog I've ever known, Grisby is my Atma, the universal soul of dog, the ideal essence, which, according to Schopenhauer's Platonic notion of true forms, exists both before and after each imperfect manifestation. Plato would disagree; he'd claim that only the *idea* of the dog is real, and Grisby is a flawed copy of the unchangeable and original essence. But how could he know? As far as I'm aware, Plato didn't have a warm bulldog on his lap, licking his knees as he wrote.

BULL'S-EYE

BULL'S-EYE IS THE dog belonging to Bill Sikes, the vicious thug in Charles Dickens's *Oliver Twist*. The dog is often assumed to be a bull terrier (the 2007 Random House Vintage edition of *Oliver Twist* has a white bull terrier on the cover, and the current Wikipedia entry for Bill Sikes claims that "he owns a bull terrier named Bull's Eye"). The source of the error is probably the 1968 musical film *Oliver!*, in which the role of Bull's-eye was played by a barrel-bodied bull terrier named Butch. In the Dickens novel, however, no breed is mentioned; Bull's-eye is described as "a white shaggy dog, with his face scratched and torn in twenty different places." Early illustrators like George Cruikshank drew Sikes's companion as a scrappy, underfed mongrel. Incidentally, the modern, long-faced bull terrier is a hybrid that didn't exist in 1838, when Dickens was writing *Oliver Twist*.

In the novel, man and dog are bound together, both victims of a cruel upbringing, both unpredictably violent. The two brutes share more than similar-sounding names; Bull's-eye has "faults of temper in common with his owner," yet the pair are inseparable, and Bull's-eye, who sleeps at Sikes's feet or by his side, is always ready to obey his master's whims. The Artful Dodger describes Bull's-eye as the "downiest of the lot" in Fagin's establishment, adding: "He wouldn't so much as bark in a witness-box for fear of committing himself; no, not if you tied him up in one and left him there without wittles for a fortnight." In return, Sikes constantly denigrates his dog, calling him a "stupid brute," a "born devil." He repays Bull's-eye's loyalty with ill-treatment, shaking him cruelly, even assaulting him with a hot poker and a clasp knife. In spite of all this, at a word or even a look from his master, Bull's-eye is ready to serve him.

Most of the time, Bill Sikes treats his girlfriend, Nancy, the way he treats his dog. In the end, he murders her in a fit of rage, with Bull's-eye as a mute witness to the crime (Bill fears the dog's bloody paw prints will "carry out new evidences of the crime into the streets"). The creature becomes a dark reminder of his master's guilt, and unable to shake the dog off his trail, Sikes attempts to drown him. Fortunately, the mutt has the sense to slink reproachfully away, eventually—and accidentally—leading the police to his master's lair. While on the run from an angry mob Sikes hangs himself; it's not clear whether his death is accidental or intentional. At the sight of his master hanging from a chimney top, in another ambiguous act of anger or possibly remorse, Bull's-eye hurls himself at the dead man's shoulders, and he, too, comes to a sorry end. Missing his aim, he lands in a ditch and, "striking his head against a stone, dashe[s] out his brains." This is how loyalty is repaid.

Bull's-eye may be the most long-suffering dog in Dickens, but he's not the only one with a brutal master. In *Little Dorrit*, the indolent Henry Gowan goes nowhere without Lion, his enormous Newfoundland. Lion is gentle and affectionate. When he encountered Gowan's fiancée, Pet Meagles, after a short absence, he "put his great paws on her arm and laid his head against her dear bosom." After Pet and Gowan are married, however, Pet learns her new husband is not only lazy but also horribly cruel. When Lion caused undue alarm, his master "seized the dog with both hands by the collar," then "felled him with a blow on the head, and standing over him, struck him many times severely with the heel of his boot, so that his mouth was presently bloody." Poor Lion is "deeply ashamed of having caused them this alarm," and in order to escape Gowan's assault, he crawls along the ground "to the feet of his mistress," but Gowan is unforgiving, kicking him over and over again until he's dead.

Like his master, Bull's-eye lives a harsh life, but that doesn't mean he's not happy. Contentment, for people as well as dogs, seems to depend largely on familiar relationships and their accustomed dynamics, however difficult they may be for outsiders to understand. We like what we know. Some dogs—at least in literature—do seem to be both deprived and content, such as the mangy dog belonging to Meursault's elderly neighbor, Salamano, in Camus's *The Stranger.* For eight years, this old man beats and insults his dog; then, every night before going to bed, rubs him tenderly with ointment for his skin disease. "He was bad-tempered," Salamano tells Meursault when his dog goes missing. "We'd have a run-in every now and then. But he was a good dog just the same."

In the case of Bill Sikes and Bull's-eye, the dog stands as a kind of avatar for the man—a common literary conceit. In Charlotte Brontë's *Jane Eyre*, the first sign of Mr. Rochester's presence is the sight of his faithful companion Pilot, a "great black and white long-haired dog" that Jane, encountering on a dark night, first mistakes for a Gytrash, "a lion-like creature with long hair and a huge head." At the end of the novel, Jane returns to find that Mr. Rochester has lost his sight in the fire that destroyed his home and can't tell who she is—but Pilot pricks up his ears when she enters the room; "then he jumped up with a yelp and a whine, and bounded toward me."

At other times, the master's dog has a more subtle function. In *Lady Chatterley's Lover*, Flossie, Oliver Mellors's spaniel, always heralds her master's approach. Before long, Lady Chatterley's heart starts to flutter whenever she sees or hears the dog. Flossie is an indicator of Mellors's presence, and stands guard during the lovers' trysts. She's a working sheepdog, a worthy companion, and such dogs, unlike the lady's lapdog, are rarely dismissed as pets or playthings. According to the French author Colette Audry, "workmen are always ready to make rude jokes about poodles and basset hounds and their doting woman owners, but wolfhounds, Alsatians, and similar breeds they take very seriously indeed."

English bulldogs are one of the breeds men take seriously, owing, presumably, to the breed's apparent toughness and stamina. In 2012, according to a survey conducted by a British men's grooming brand, the English bulldog was voted "the manliest dog on the planet." It's hardly surprising, then, that all kinds of macho objects and activities should be named after the sturdy-looking creature. There are Bulldog jeans and Bulldog knives; there's Bulldog Gin, Bulldog Hot Sauce, and Bulldog Hardware. There are hundreds of sports teams named the Bulldogs. Vehicles named after the breed include a British fighter aircraft, a Royal Navy ship, an armored personnel carrier, and a German tractor. Could any dog be more butch?

Ironically, the English bulldog is a rather delicate beast, docile and affectionate, prone to health problems and easily tired. His small French cousin, on the other hand, although culturally coded as feminine (see ISSA), is muscular, dominant, and tough as a little tank, not to mention stubborn. Grisby was not the most obstinate dog in his obedience class—that dubious honor went to a terminally intractable terrier whom everyone, including his genteel owner, referred to as "the Nazi"—but he certainly placed a close second. Sometimes he did what was asked of him, but his "training" took only until we got home, whereupon he'd jump out of the car, barge rudely ahead, push through the front door, and run into the house. As our obedience instructor kept reminding us, going to class is the easy part; the hard part is reinforcing the lessons at home. She was, I thought, infinitely patient and, I was pleased to find, had no beef with affection and rewards. At first, I was worried she might endorse the techniques promoted by Cesar Millan, who insists that we assert dominance over our dogs instead of treating them like babies.

Personally, I don't believe you need to act like an alpha dog at home, nor do I think badly trained dogs will always try to assert themselves over strangers. Still, I do understand the importance of consistency, and I realize Grisby is sometimes disobedient because, unable to bring myself to punish him, I've been unpredictable in my demands and rewards. In this respect, David has been—and continues to be—the better master. He's firm, consistent, and not afraid to lay down the law. When—as sometimes happens—Grisby slips out of our apartment and runs into the hall, one strong word from David can make him skid to a halt, lower his ears, and submit to the leash. If I'm the one reprimanding him, however, he keeps still until I approach, then jumps up like a jack-in-the-box and runs off, throwing me a backward glance that says, "So long, sucker!"

In this way, perhaps, David's relationship with Grisby is healthier than mine. With me, Grisby is enmeshed; with David, he knows his place. In other words, David has what most people would probably consider to be an appropriate kind of relationship with his dog. He loves Grisby, worries when he's sick, enjoys having him around, but doesn't miss him—doesn't even think about him, doesn't even really notice—when he's not there. He has his own pet names for Grisby—Bright Eyes, Big Boy, Señor—that are affectionate but not infantilizing. It seems ridiculous for me to be jealous, but sometimes I wonder whether, as males, David and Grisby have a bond I'll never share. It's tempting to romanticize the man-dog connection, and to overlook the fact that it can be instrumental or exploitative, or that it usually involves questions of aggression and control.

These issues appear most overtly in the hypermale world of dogfighting, a practice that goes back to ancient times. The Egyptians, Greeks, and Babylonians all employed fighting dogs on the battlefield. During the Roman invasion of Britain, the conquering legions were impressed by what early historians referred to as the *pugnaces britanniae*: the fighting dogs of Britain. The specific breed of these ferocious, battle-ready beasts is unknown, but in light of an early reference to them as "broadmouthed," it's widely believed they were remote ancestors of the modern-day mastiff.

Soon after their invasion, the Romans began to import British fighting dogs, even appointing an officer whose job was to select especially pugnacious animals to send abroad. Some were trained to fight in battle; others were turned into gladiators and pitted against bulls, bears, and wild elephants in the Colosseum, a precursor to modern bullfighting. Later, the *pugnaces britanniae* were used in bearbaiting, a "sport" that flourished in the sixteenth century and was especially popular among English noblemen (ironically, it's the blue bloods who pursue blood sports most earnestly). By the early nineteenth century, the pastime had become less common, owing to the increasing scarcity and rising cost of bears (as well as growing concerns about cruelty to animals), and in 1835 bear- and bullbaiting were both outlawed by an Act of Parliament. Henceforth, these "sports" were replaced by the cheaper, legal alternative of dog-on-dog combat, and fighting breeds were crossbred to create agile and vicious creatures capable of brawling for hours at a time.

Shortly before the American Civil War, English fighting dogs were imported to the United States, where they were mated with native breeds. Dogfighting quickly became a popular spectator and betting sport in the United States, and the United Kennel Club created formal rules and sanctioned referees. Fights were held in taverns and halls, and railroads would sometimes offer special fares to passengers traveling to well-publicized events. The observer of a Brooklyn dogfight in 1876 described its spectators as a "villainous-looking set . . . more inhuman in appearance than the dogs . . . a crowd of brutal wretches whose conduct stamps them as beneath the struggling beasts." Unsurprisingly, perhaps, most dogfighters were men in typically macho working-class professions: police officers, soldiers, and firefighters. When dogfighting became illegal in the 1930s and '40s, it was driven underground, where it continues to thrive, despite its being classed as a felony in all fifty states.

"Let dogs delight to bark and bite," begins a hymn by the English theologian Isaac Watts. This is the line usually taken by defenders of legalized dogfighting—that dogs naturally exult in their strength and are eager for combat; that fighting, in other words, is "in their nature." I know there are fighting rings in Baltimore, and I sometimes worry Grisby might be stolen for use as bait. Pet theft is apparently on the rise in the city, though since it's lumped in with other kinds of property theft, it's difficult to know how widespread it really is.

Such theft is certainly not as common as it was in nineteenth-century London, when substantial ransoms would be asked for the animals' safe return. The most notorious of these mercenary pet pilferers was a gang whose members called themselves "the Fancy." Their modus operandi was to wait until the dog was momentarily unattended, lure the unsuspecting creature—usually with liver mixed with myrrh or opium, or sometimes with a bitch in heat—then shove the poor animal in a sack and disappear into the crowd. When the gang's demands weren't met, the dog's paws or even its head would be delivered to its owner. Flush, Elizabeth Barrett Browning's spaniel, was kidnapped three times by the Fancy, and each time she unhesitatingly and immediately paid the ransom (see FLUSH). Who can blame her?

Still, when I asked an animal control officer whether I was taking a risk by leaving Grisby tied up outside a Starbucks, he looked amused. "No risk at all," he assured me, condescendingly. "Anybody that's involved in illegal activities is going to want to stay under the radar as much as possible. If they wanted dogs as bait, they're not going to steal one off the street. For one thing, you can just go and get a mutt from the pound—this city's full of people trying to get rid of dogs they can't afford to keep. Another thing—if you steal a purebred, it's probably going to have a microchip and it's going to be worth some money, which bumps it up from a theft to a felony. Nobody's going to take those kinds of risks for what's at stake."

I felt foolish. When you think about it, the idea of gangsters emerging from the ghetto to steal "our" innocent pets is really absurd; what's more, it bespeaks all kinds of race and class anxieties. These sensitive issues also saturate the discourse around pit bull "rescue" campaigns, in which dogs are taken from young black men in the city's run-down neighborhoods, inoculated, bathed, "altered," given friendly names, adopted by middle-class families, and taken to live in the suburbs. We do to the dogs what we really want to do to the barbarians who breed them: make them submit.

CAESAR III

CAESAR III IS a Boston terrier who appears in the short story "Coming, Aphrodite!" by Willa Cather (first published in August 1920 under the title "Coming, Eden Bower!"). The narrative's central character is Caesar's master, Don Hedger, a solitary artist whose ascetic life is thrown into turmoil by the arrival of a glamorous new resident to the Washington Square boardinghouse in which he lives. The sensual Eden (real name: Edna) Bower is a singer who uses her looks and talent to draw the crowds.

When we first meet them, Don and Caesar are living a quiet, uneventful life in Hedger's small studio. Caesar, set in his ways, is a grouchy and sullen creature with an "ugly but sensitive face." People complain about the dog's surly disposition, but Don explains that it's not Caesar's fault—"he had been bred to the point where it told on his nerves." Every day, the pair follow the same quiet, austere routine. In the morning, Hedger gives Caesar a bath in the rooming house's shared tub and then rubs him into a glow with a heavy towel. All day, Don paints, and Caesar sits alertly at his feet; in the evening, the pair eat together at the same basement oyster house. For days on end, Don talks to "nobody but his dog and the janitress and the lame oysterman." In summer, when the nights are hot, Hedger climbs up a ladder to the roof, carrying Caesar under his arm, and they sleep together side by side under the stars.

Eden Bower first appears in the hall outside the neighbors' shared bathroom.

"I wish you wouldn't wash your dog in the tub," she complains to Don.

Until then, "it had never occurred to Hedger that anyone would mind using the tub after Caesar," but suddenly made ashamed by Eden's dignified beauty, "he realized the unfitness of it." Eden Bower, he immediately realizes, is a different kind of creature from males like Caesar and himself. Listening to her sing and play the piano, Hedger finds her mesmerizing. He discovers a crack in his studio wall and starts to spy on Eden every morning when she exercises in the nude. Finally, he gets up the courage to ask her if she'd like to join him on a trip to Coney Island. Eden considers the prospect; her doubt focuses not on Don but on his dog. She concedes, but only as long as Caesar is left behind.

Hedger is taken aback. "But he's half the fun," he argues. "You'd like to hear him bark at the waves when they come in."

Eden knows better. "No, I wouldn't," she retorts. "He's jealous and disagreeable if he sees you talking to anyone else."

So Caesar is left behind, "lying on his pallet, with a bone" while the couple spend the day at Coney Island. Here, Eden finds herself growing attracted to Hedger, though she's slightly afraid of his brutality ("she had often told herself that his lean, big-boned lower jaw was like his bull-dog's").

When they return to Washington Square, Eden and Hedger become lovers, and decide to open up the double doors that separate their rooms in the boardinghouse. All at once, Hedger's dark, cave-like lair becomes a bright love nest, and "Caesar, lying on his bed in the dark corner," is startled by this invasion of sunlight: "the side of his room was broken open, and his whole world shattered by change." A miserable interlude passes—miserable, at least, for the dog—during which Don bestows all his attention and affection on Eden Bower. True to his name, Caesar III is put in third place. Before long, however, Eden has become such a hit in New York that she's booked on a European tour, summoned to take up her place in a feminine world of fashion and glamour. She soon becomes wealthy and widely known, while Hedger, the serious artist, remains alone in his creative struggle, uninterested in the wider world. He closes the doors on the sunlight, returning to his quiet life with his dog.

I'm so fond of Caesar that it disappoints me to discover everyone who's written about this story regards him either as a symbol—invariably "phallic"—of Don Hedger's masculinity, or as a representation of his master's artistic practice (unfriendly, inward-looking, a force that prevents wider engagement in the world). I wonder: Why does Caesar have to represent anything? Why can't he just be Caesar? To me, "Coming, Aphrodite!" is the story of a tormented love triangle that develops between Hedger, Caesar, and Eden Bower, whose arrival upsets the perfect balance of man and dog.

I'm drawn to Caesar in part because he's a Boston terrier, and bully breeds remind me of Grisby, but also because he's such a fully realized character in the story—all the more reason why it seems wrongheaded to see him as a mere symbol. Like Don Hedger and Eden Bower, Caesar III has his own scent, texture, and personality. In what to me are the story's most moving scenes, the narrative voice slides almost imperceptibly from

Don's perspective to Caesar's, and though we may not consciously notice it, we're experiencing events from Caesar's point of view. When Don carries his dog up the ladder to the roof, for example, we learn, of Caesar, that "never did he feel so much his master's greatness and his own dependence upon him, as when he crept under his arm for this perilous ascent." The roof to Caesar is "a kind of Heaven, which no-one was strong enough to reach but his great, paint-smelling master," where he and Don lie together under the stars, feeling the same delight, "lost in watching the glittering game."

I couldn't help thinking of Caesar when I came across a bull terrier in Edith Wharton's novel *The House of Mirth.* In an early scene, Mrs. Bertha Dorset gets on board a train "accompanied by a maid, a bull-terrier, and a footman staggering under a load of bags and dressing-cases." This dog at once caught my attention, yet oddly enough, we never see or hear of the creature again. Bertha Dorset is a major character in *The House of Mirth*, and Edith Wharton has a fine eye for detail, so it's surprising to come across this elusive bull terrier. Perhaps Mrs. Dorset borrowed the dog, regarding it as a fashionable accessory for a trip to Bellomont, the country home of her friends Judy and Gus Trenor. She does have a calculating, opportunistic nature, and she might very well forget, lose, or give away a dog that had lost its charm. And we're never informed that the bull terrier belongs to Bertha Dorset—it could be a gift for one of the Trenors' two teenage daughters, or for another member of the house party. It might even belong to her maid.

However unlikely it seems that a maid would be permitted to travel with her bull terrier, it makes more sense than the notion that Wharton would have introduced a dog into her novel simply to forget about it in the next chapter. From her childhood, dogs were her most loyal and lasting companions, though she preferred smaller breeds than bull terriers. In a well-known

photograph, she has a Pekingese on each shoulder, the two creatures nestling comfortably atop the leg-of-mutton sleeves of their
mistress's dress. In other images, the author sits with dogs on her
lap, under her arms, or at her feet. In one or two pictures, she sits
at a desk, pen in hand, but these images were posed; Wharton
actually wrote in bed, with her coffee, the morning papers, and
her dogs all spread out around her. Like many women, she was
not ashamed to admit she slept with her dogs. According to an
article in the journal *Emerging Infectious Diseases*, 25 percent of
female dog owners confessed to doing the same (as opposed to 16
percent of men; evidently, men are either more fastidious, less
affectionate, or simply lying).

Wharton's first dog was a puppy named Foxy, bought for her
by her father when she was a young girl. "How I loved that first
'Foxy' of mine, how I cherished and yearned over and understood
him!" wrote Wharton as an adult, looking back on her childhood.
"And how quickly he relegated all dolls and other inanimate toys
to the region of my everlasting indifference!" The author of *The
Age of Innocence* felt she shared a special understanding with
animals, writing that she was one of the few who "love and understand the little four-foots . . . [and] have the mysterious animal affinity [to] communicate with [them]." Her relationship
with Foxy, she wrote, "made me into a conscious sentient person,
fiercely possessive, anxiously watchful, and woke in me that long
ache of pity for animals, and for all inarticulate beings, which
nothing has ever stilled."

The House of Mirth was written in 1905, when it was presumably acceptable for a lady to bring a small dog with her—or
with her maid, at least—on the train. These days, Amtrak has a
strict no-pet policy (with an exception for service dogs), though
the inner-city rail networks in Boston, San Francisco, and Seattle
all allow well-behaved dogs to ride with their owners. Outside
the United States, railways are more accommodating. Dogs are

allowed on trains in much of Europe. We took Grisby with us to London a couple of years ago, and he sat beside me on buses and trains, even in taxis. He remained composed and unflustered even in a subway car packed with patriotic revelers on their way to celebrate the Queen's Diamond Jubilee (Grisby may not be a British bulldog, but he knows how to keep a stiff upper lip).

Of course, the best way to get to know a city is by walking, and this is true of dogs as much as humans. I get a lot more exercise since Grisby's arrival in my life; these days, I prefer to travel by foot just for the pleasure of watching him trot along by my side. I like the way he draws my attention to things I'd never noticed before: stains on the sidewalk, discarded food, chewing gum, feathers, cigarette butts. For Grisby, every season has its special delights. In winter, he'll climb through the snow, catching flakes in his mouth. In spring, he'll stick his snout into damp gutters and rain-soaked trash piles. In summer, he, like Caesar, loves to bark at the waves on the beach, and in fall, he cocks his leg against piles of dried leaves; then, when he's covered them with his smell, he'll kick them all over the street, just to show everyone who's boss. The only weather he really dislikes is rain. In wet conditions, his walk will turn into a run as he charges around the block at top speed, running up every stoop, trying to dodge inside every doorway, as if hoping some sympathetic stranger will take pity on him and invite him to dry off by the fire.

He enjoys getting dirty, but unlike Caesar III, he doesn't enjoy his baths, and for this reason I try to keep them limited to once a month. It's hard to resist, though, because it's such a treat to have him in the tub with me, to hold his warm, wet body firmly between my knees. He soon stops struggling, allowing me to wash his ears gently, remove the dirt from his wrinkles, shampoo his neck, back, and belly. After rinsing him off, I'll lift him out of the tub and watch him run around the room, shaking his body dry, looking oddly naked without his collar.

Grisby loves to walk, but he likes other modes of transport as well; he enjoys riding in the small red cart used by the concierge in our building to deliver large packages. If he loves to be pushed, he loves even more to be driven. Like Mr. Toad, he has a head for motoring adventures, and he'll start racing in excitement whenever he realizes we're heading for the garage. Like all dogs, he loves sticking his nose out of the window and catching a breeze, letting his ears blow back in the wind. He's even worked out how to press the electric window button with his paw (though "worked out" may not be exactly the right way of putting it), which, on a cold morning, can lead to some frustrating battles between us (of course, he can undo the window lock, too).

Best of all are our rides back from the beach together on summer days, when Grisby—warm and wet, with sand in his fur—is strapped into his harness on my lap. On these drives, we're connected at a physical level, like Don and Caesar climbing up to the roof. Strapped to my body, Grisby presses heavily into my stomach, and our bodies respond together to the jolts of the car. At those moments, with a wet bulldog snoring on my lap, it's as though we're merged together organically, a hybrid creature of flesh and fur, a single animal with two beating hearts.

[4]

DOUCHKA

ÐOUCHKA, A TROUBLESOME and neurotic German shepherd, was the subject of *Behind the Bathtub*, a book that won the Prix Médicis (a major French literary award) in 1962. The author of this sober and touching memoir was Colette Audry, a French literary critic, screenwriter, and expert on the work of Jean-Paul Sartre, with whom she often collaborated. She was a militant feminist, deeply involved in the anti-Stalinist left.

When she first acquires Douchka, Madame Audry is divorced and living in Paris with her teenage son. The dog's parents, she discovers too late, were brother and sister, and as a result Douchka has various psychological problems, the most serious of which, from Audry's perspective, is her furious barking in cars. In the company of Douchka, any trip, however short, becomes a nightmare. Sedatives are completely useless. So relentless and unbearable is the racket she makes that at one point Audry seriously considers having the dog's vocal cords removed. The barking gets so infuriating that she often wants to throw Douchka bodily out of the car, and on one occasion actually opens the door and lets the dog fall into the road, leaving the desperate creature to run for two miles on swollen paws—a punishment the dog's mistress bitterly regrets.

Barking is Douchka's worst problem, but not her only one; in fact, it may not be going too far to describe the dog as barking mad. She's nervous and needy and can't be left alone, demanding Audry's constant attention, dragging her away from her writing and political activism. When her mistress goes out at night to put up antigovernment posters in the Paris streets, Douchka manages to escape and follows her; for the activists, the dog's anxious barking becomes a dangerous liability. As time passes, Douchka's needs gradually compel Audry to give up most of her customary activities, and often prevent her from leaving her apartment. On top of this, Audry is convinced it would be wrong to "alter" her dog, and consequently, twice a year, she ends up "fighting off Douchka's would-be admirers like an officious chaperon."

As she gets older, Douchka grows increasingly disturbed, and Audry finds herself almost overwhelmed by the responsibility of caring for the creature. For a while, she considers putting Douchka to "sleep," but is unable to go through with it, and she finally realizes that something must give. "I could neither cure Douchka or her neurosis," she eventually admits, "nor myself of the enormous emotional burden she laid on my life." For Douchka's sake, then, Audry gives up "what no man had ever taken from me—my freedom of movement and decision," and accepts the kinds of restrictions that, as a militant feminist, she's battled against all her life. Yet once she's stopped struggling, Douchka's mistress starts to find that although in some ways her independence has been curtailed, the payoff is unexpectedly sweet. Now she can devote herself completely to the intractable Douchka, and she confesses that "loving her gave me a special pleasure: it was unlike anything else I have ever experienced, a mixture of responsibility, amusement, and gaiety, a small deep-rooted delight concentrated on her and her alone."

Reading *Behind the Bathtub* is a mixed experience. The book is beautifully written but often very sad, and Douchka can be infuriating. It made me realize how blessed I am by Grisby's placid nature and traveling chops. Sure, he likes to be around me, even to the extent of following me to the toilet and pushing open the bathroom door with his flat snout, but he's never too clingy. At the beach, I'll lie on a blanket and read while he plays nearby like a well-behaved child, paddling and exploring, safe in the knowledge that, should he need me, I'll always be close by. When we walk in the woods, he'll trot at my feet, but will fall back if he finds something interesting to sniff or chew. I don't slacken my pace—I know that, before long, I'll hear him panting and snorting behind me as he runs to catch up. In fact, when we're apart, I'm sure I suffer more than he does, missing all the little signs of his presence—his small sighs and grunts, the sound of his claws on the floorboards, his jingling collar, his soft ears rubbing against my knees.

I try not to, but I often find myself wondering what he's feeling in my absence, which, in J. R. Ackerley's novel *We Think the World of You*, is the first step down the slope to madness and heartbreak. In this book, the narrator, Frank, upsets himself by worrying about the dog owned by his young lover, Johnny, who's serving time in prison. Johnny's German shepherd, Evie, is being "cared for" by the young man's working-class family, whose treatment of her—she's left alone in a small courtyard for ten hours a day—strikes Frank as profoundly cruel. He drives himself right to the edge of a nervous breakdown imagining the dog left at home alone, "hope constantly springing, constantly dashed." He pictures how "she would gaze longingly at the lead on the wall, go over to it to investigate it with her black nose, employ all her little arts to draw attention to her needs, and get nothing, nothing . . . Day after day, day after day, nothing, nothing; the giving and the never getting; the hoping and the waiting for something that never comes."

"I—I can't bear to think of her," Frank confesses to Johnny one day, during a prison visit. "Her loneliness. I can't bear it. It upsets me." His suffering is made worse by the fact that every time he gets up to leave after visiting Evie, she becomes hysterical, jumping up and down and looking at him with desperate hope. "It always affected me with a sensation of hysteria similar perhaps to her own," says Frank, "a feeling that if I did not take care I should begin to laugh, or to cry, or possibly to bark, and never be able to stop." Even after he's adopted the dog and taken her into his home, Frank still worries about her during the day, when he's at work and she's at home alone. "That she was awaiting my return I had no doubt at all," he says. "I knew that she loved me and listened for me, that whenever a knock came at the door her tall, shell-like ears strained forward with the hope 'Is it he?'" In a similar fashion, Thomas Mann puts himself in the mind of his setter, Bashan. When Mann leaves for work every morning in the city, he confesses that "a pang goes through my heart—I mount the train with an uneasy conscience. He has waited so long and so patiently—and who does not know what torture waiting can be! His whole life is nothing but waiting—for the next walk in the open—and this waiting begins as soon as he has rested after his last run."

If Douchka, Evie, or Bashan were human beings waiting anxiously all day for one person to come home, we'd probably describe them as being "in love," perhaps even to an obsessive degree. But is this kind of love the same as human love? Marjorie Garber, in her book on the subject, makes the case that "dog love is local love, passionate, often unmediated, virtually always reciprocated, fulfilling, manageable. Love for human beings is harder. Human beauty and grace are fitfully encountered: a child grows up and grows away, a lover becomes familiar, known, imperfect, taken for granted." Our complicated bond with our dogs, argues Caroline Knapp in *Pack of Two*, is profoundly gratifying because "dogs occupy the niche between our fantasies about intimacy and our more practical, realistic needs in relation to others, our needs for boundaries and autonomy and distance."

Dogs know instinctively how to show their feelings for us, but it's hard to know how to love them back. Some dog owners spend thousands of dollars on designer doghouses; some ruin their pets' health with too many treats; some take their pals to sheepherding boot camps, or run them through agility trials every weekend. I do none of these things; I simply love to be with Grisby. I love to kiss and pet him, but while he seems to understand the point of my affection, he doesn't always appreciate getting it as much as I enjoy giving it. This often makes me feel a little Humbert Humbert–ish, especially when Grisby's sitting on my lap in the car and I have access to parts of his body that are normally inaccessible to me, like his soft piebald underbelly. Should I feel ashamed of myself?

The question remains: Do dogs "fall in love" with us the way we do with them? According to John Bradshaw, the author of *In Defense of Dogs*, the experience isn't exactly the same. When a dog licks your face, says Bradshaw, it's gathering information about you from your breath and sweat glands, learning when you had your last meal and whether there might be any bits of it left over. This is a gesture it's instinctively programmed to go through every time it comes close to another friendly mouth, whether human or canine. In other words, when Grisby nuzzles my face, he's displaying not affection but the same kind of instinctive curiosity that leads him to sniff a drain or stick his nose into the trash, though—from his point of view at least—my face is rarely as rich or rewarding. "Dogs are obviously attached to their owners—in the sense of their behavior, in the sense that they follow them around," Bradshaw concedes, finally getting to the crucial question: Does your dog actually love you? At last, he gives me the answer I've been dying to hear ("Of course it does!"), but it's too glib for me to take seriously. It's too easy, too ingratiating: a sop to Cerberus. Dog love is more complicated than that.

After all, whatever word we might choose to describe them, our feelings for our dogs—and their feelings for us—may be gratifying, but they can also be painful and tormented, even more than our emotions for other human beings. In *We Think the World of You*, Frank is devoted to Evie, but his dedication is selfish, peevish, and sometimes even toxic. "I loved her; I wished her forever happy," he admits, "but I could not bear to lose her. I could not bear even to share her. She was my true love and I wanted her all to myself." As this novel testifies, love for dogs can be confusing, contrary, and full of terrible suffering.

Like many dog books, *Behind the Bathtub* ends unhappily. While she's still quite young, Douchka is bitten by an infected rat, grows sick, and one night crawls to the place where she always used to hide when she'd been reprimanded—behind the bathtub. It's a heartbreaking scene. Audry describes it clearly and without sentiment, struggling to understand the unique nature of the wordless, cross-species relationship the two females have shared. Reading this part of the book, I found myself wondering whether, when Grisby dies, I'll look on our relationship with sorrow and regret, or whether memories of him will fade away fast as I move on to my second dog. Perhaps I'll look at my photographs of Grisby the way I look at pictures of former lovers, wondering, as I toss them into the trash, what I ever saw in him, and how I could have deluded myself for so long.

The Prix Médicis is traditionally awarded to underrated authors with the aim of boosting their literary reputations and launching them to the next level of their careers (prizewinners have included such acclaimed writers as Monique Wittig, Elie Wiesel, Hélène Cixous, and Bernard-Henri Lévy). Yet the prize did little for Colette Audry, and some even ridiculed the committee for giving this important award to "a dog book." Sadly, this brave and serious work of autobiography has long been out of print; its critical reception is illustrative of our uneasy cultural relationship not only with dogs but also with those who write about them. Many French reviewers were unable to take *Behind the Bathtub* seriously, dismissing it as a minor or negligible work.

Despite England's reputation as a nation of animal lovers, response to the British edition (*Douchka—The Story of a Dog*) was equally condescending. According to one Sister Mary William in *Best Sellers*: "The book is easy reading and, I suppose, pleasant reading for dog lovers." Sister William added sneeringly: "It seems to me that five dollars is quite a price to pay for this sentimental journey into the past of dog and mistress even though it . . . is said to have been a best seller in France." Naomi Lewis in the *Observer* called *Douchka* "an angry, tormented book," and Robert Nye in the *Guardian* described it as "run-of-the-mill animal stuff, all right if you can stomach it but otherwise about as appetising as cat's meat and dog's biscuits." Francis Wyndham, reviewing *Douchka* in the *New Statesman* in 1963, was of the opinion that Audry "presumably writes that aggressively colloquial prose often favoured by intellectual French women" and suggests that "books about animals often seem unduly egotistic" because "there isn't that much that *can* be written about them." Finally, in her 1990 obituary of Colette Audry, Maryvonne Grellier in the *Guardian* completely mischaracterized *Behind the Bathtub* as "based on a childhood episode when she found the family's pet dog dead behind the bath."

We still make fun of women like Colette Audry who love their dogs "excessively." (But who decides how much love is "too much"?) There seems to be an unstated assumption that this love is being "wasted" on an animal, that women who devote themselves to their dogs are slightly unhinged. For this reason, many women keep such feelings to themselves—but such emotions may be far more common than we think. The website Dogster includes a regular feature called "Doghouse Confessional," where readers send in their secrets about their dogs. Past columns (all by women) have included "I Love My Dog More Than I Love My Husband," "I Put My Dog's Happiness First," and "My Dog Has Outlasted All My Romantic Relationships." If such feelings are widespread, moreover, it should be cause for celebration, not concern. You might love your dog more than you love your husband, but loving a dog doesn't mean you stop loving people; in fact, evidence suggests that love for animals encourages a broader sense of general empathy.

Even today, despite the increasing importance of dogs in our lives, books about them are invariably dismissed as sentimental and lighthearted, lucrative but simplistic, the lowest form of literature. Alice A. Kuzniar, the author of *Melancholia's Dog*, opens her thoughtful book on human-canine kinships by remarking that the subject of dogs is presumed to be unfit for serious scholarly investigation; "it is held," she writes, "to be sentimental, popular, and trivial. . . . Whenever I had to explain and justify to what I was devoting years of research and writing, I felt embarrassed." Why can't we let ourselves take dog love seriously? Is it because, if we did, we'd have to think seriously about other nonhuman animals, including those on our dinner plates? One way to keep these anxieties at a distance is to make fun of people who've got their pets out of all proportion; this is how we can restore the balance, reassuring ourselves that of course, although some people take their feelings for dogs too far, *we* know dog love isn't "real love" (if it were, what would stop us from choosing dogs over people?). This, at least, is the only way I can possibly make sense of one reviewer's perplexing summary of *Behind the Bathtub*: "Beneath the story of Mme. Audry and Douchka lies, almost hidden, the terrible tragedy of a loveless life."

EOS

EOS, NAMED AFTER the Greek goddess of dawn, was a beloved female greyhound belonging to Prince Albert, the husband of Queen Victoria. The German-born prince had a lonely childhood. His parents' marriage was notoriously turbulent, and they divorced when Albert was five. The prince's mother was exiled from court, his father grew cold and distant, and the boy's only companion was his brother, Ernest, but when Albert was fourteen, Ernest left for college in Bonn. To keep the young Albert entertained, Eos, then a six-month-old puppy, was delivered to the royal family's home in the Schloss Rosenau, and the pair were soon inseparable. "She was my companion from my fourteenth to my twenty-fifth year," wrote the prince upon his greyhound's death, "a symbol therefore of the best and the fairest section of my life."

When it was Albert's turn to go to school in Bonn, Eos went with him, and when the prince married Victoria in 1840, at age twenty, the greyhound was sent ahead to Buckingham Palace in the care of her own personal valet. She'd grown into a fetching beast, black with white paws, a white underbelly, and a white tail tip. The newlyweds were both fond of animals and kept a number of dogs at court, but most of these belonged to the queen; Eos alone was decisively Albert's dog. Six months after their marriage, the prince turned twenty-one, and in honor of the occasion, Victoria commissioned the royal jeweler, Garrard, to make an eight-by-ten-inch silver model of her husband's greyhound (according to her diary, Albert was "much pleased" with his gift).

On their first anniversary, the queen presented Albert with another surprise: a portrait of Eos by the painter of animals Sir Edwin Landseer. This famous picture is entitled *Eos, A Favourite Greyhound, Property of HRH Prince Albert*, and it shows the dog posed against a rich red tablecloth, her muscles tense and rippling. Around her neck is a royal red-and-gold collar. With an eager and servile expression, she guards her master's top hat and gloves, which are laid out ready for him on a hoof-footed, deerskin-covered stool. On a table behind the dog lies her master's cane, topped with a red tassel and an ivory handle.

In terms of its pose, content, and composition, *Eos, A Favourite Greyhound* is regarded as an early example of the Victorian tradition of formal pet portraiture, but although the title describes it as a painting of Eos, it is, essentially, a portrait of Albert in absentia. The prince's status and dignity are represented by his property: the eager bitch and the gentlemanly accessories. In light of the fact that Albert was himself sometimes dismissed as Victoria's lapdog, perhaps the queen intended the painting as a form of subtle compensation. Either way, Albert was not offended; in fact, the portrait was so well received that she chose to commission another painting by Landseer for Albert's next birthday, in 1842.

In this picture, entitled *Victoria, Princess Royal, with Eos*, the new princess lies in her crib, and the greyhound—the prince's other daughter—rests her slim snout protectively between the child's bare feet. The dog's gentle pose is perhaps deliberately reminiscent of the legendary hound Gelert, who gave his life to protect a royal baby (see HACHIKŌ).

Albert himself liked to paint and also dabbled in design. One of his more elaborate creations was commissioned over the winter of 1842–43, again in conjunction with Garrard: a colossal set of three gilded candelabra. The main part of this enormous ornament is a four-foot column whose base is decorated with carvings of Eos and three of Queen Victoria's dogs: her Skye terrier Islay ("a darling little fellow, yellow brindled, rough long hair, very short legs and a large, long, intelligent good face"), her Scottish terrier Cairnach ("he had such dear engaging ways"), and her favorite dachshund, Waldmann. This grotesque gewgaw may have been too much even for Queen Victoria; the following year, it was unloaded on Viscount Melbourne, the former prime minister, on the occasion of his retirement, and it was known thereafter as the Melbourne Centerpiece.

From her portraits and sculptures, we know Eos was a sleek and dignified dog, but we know little about her personality or temperament other than a brief description included by Albert in a letter he wrote to Victoria from Germany before their marriage. In this letter, he describes Eos as "very friendly if there is plum-cake in the room . . . keen on hunting, sleepy after it, always proud and contemptuous of other dogs." She may have been rather less keen on hunting after January 1842, when she was accidentally shot by Prince Ferdinand, a relative of Albert's visiting from Germany. ("Favorites often get shot," Lord Melbourne reassured the queen, adding that he "has known it happen often.")

When the queen informed her uncle, King Leopold of Belgium, of the accident, he immediately replied to say how sorry he was to hear about what had happened to "dear Eos, a great friend of mine," and expressed his annoyance at the culprit, Prince Ferdinand, suggesting that "he ought rather to have shot somebody else of the family." The queen's subsequent letters are full of information about the dog's recovery and convalescence. On February 1, 1842, Victoria wrote that Eos "is going on well, but slowly, and still makes us rather anxious." Four days later, she wrote to let her uncle know that "Eos is quite convalescent; she walks about wrapped up in flannel."

Happily, the greyhound went on to live two more years after her injury, dying in 1844. Her demise may have been hastened by overindulgence. Shortly before the dog's death, Victoria wrote to tell King Leopold that Eos had recently suffered from an "attack" that was attributed to "overeating (she steals wherever she can get anything), living in too warm rooms, and getting too little exercise." She died very suddenly, after seeming quite well an hour before. Albert was devastated. "I am sure," he wrote to his grandmother, "you will share my sorrow at this loss. She was a singularly clever creature, and had been for eleven years faithfully devoted to me." Eos was buried beneath a mound above the slopes at Windsor Castle, the resting place of many royal pets before her. On hearing of her demise, Viscount Melbourne, perhaps alarmed that more hideous mementos might be forthcoming, declared himself "in despair at hearing of poor Eos," but he was on safe ground: the Prince Consort limited himself to designing a life-size bronze monument marking the spot of the greyhound's grave, and a second sculpture based on Landseer's original portrait. "Poor dear Albert," the queen wrote in her journal. "He feels it terribly, & I grieve so for him." Grieving was something Victoria did well; her mourning for Albert, after he died in 1861, lasted almost forty years.

Even today, dogs are injured and sometimes killed in hunting accidents, though far fewer than in Albert's day, thank goodness.

Still, every age brings its own dangers, and whatever other concerns he may have had about Eos, at least Albert didn't have to worry about her suffocating in the backseat of a car, overheating in an airplane cargo compartment, or drowning in a swimming pool—all accidents to which, as I know only too well, bulldogs are especially prone. The last of these dangers is perhaps my greatest anxiety, at least in relation to Grisby. It's obvious these top-heavy dogs aren't natural swimmers. Their bodies don't bend in the middle, and their legs can't paddle fast enough to keep them afloat. "Do not allow your bulldog near water!" warns the author of *How to Raise and Train a French Bulldog.* "He will sink like a stone!"

But dogs are individuals just as people are, and the fact is, some bulldogs love to swim, just as others—as YouTube amply testifies—love to skateboard, dive, ride on the back of motorcycles, and rock out to the blues. Grisby happens to be one of those dogs that just love water (though he doesn't like rain and flees from the tub). The first time we took him to a beach, he ran straight into the sea and began to do an odd kind of doggy paddle. Despite his manifest enthusiasm, however, he's too heavy to keep himself above the surface for long, and I'm always worried about him getting in over his depth.

For two years we lived in California, renting a beautiful ranch-style house with a pool in the backyard. At first, I was kept awake at night by the thought of coming home from work only to find my bulldog's dead body at the bottom of the pool. And for a while, it seemed my fears were justified—on two occasions, after watching us dive into the deep end, Grisby, wanting in on the fun, did the same thing. Luckily, these two soakings seemed to work as shock therapy; after that, he seemed afraid of the pool and, apart from occasionally growling at the floats, wouldn't go near it. I couldn't even get him to sit in the shallow end to cool off. He preferred to snooze under my deck chair in the shade, and his happy snores gave me a good reason to stay lazing in the sun. (One of the many advantages of having a dog is that it gives you an excuse for doing things that might seem too indulgent if you were doing them alone. "I don't like to disturb the dog," you can tell yourself as you linger another hour in bed or at the beach.)

Back home on the East Coast, I take Grisby to the Chesapeake Bay to escape the summer heat. Our favorite beach is just a twenty-minute drive from the city; it's free, easily accessible, and totally deserted. Here's the catch: it's filthy. For most people, this would be a deal breaker, but dogs are different. It's true that the water is cloudy and the sand strewn with plastic debris, but none of that bothers Grisby, so it doesn't bother me. One of the many things he's reminded me of is that it's more fun to be dirty than clean. He loves the smells and textures of trash, and he spends hours digging for buried chicken bones and discarded sandwiches. We spend many summer mornings on our special dirty beach, sunbathing, swimming together, dozing, and searching for treasure. At times, I have to veto Grisby's playthings—used diapers and dead fish are going too far—but in short, I've learned to love a beach with dog-friendly detritus.

Speaking on behalf of Grisby, filth is fun, and it's a rare dog that looks forward to bath night. Charles Robert Leslie, a nineteenth-century royal academician, wrote in his autobiography that when Victoria returned from her coronation, she heard her spaniel barking in the hall, and was apparently "in a hurry to lay aside the sceptre and ball she carried in her hands, and take off the crown and robes, to go and wash little Dash." It's hard to imagine the older Victoria washing her many dogs, and equally difficult to imagine the dignified Prince Albert bathing Eos. On the other hand, I can't imagine the stately Eos bounding, as Grisby did recently, into a stagnant cemetery pond and emerging covered in stinking graveyard mud (naturally, he climbed into my lap in the car and sat there steaming and reeking all the way home). I also can't help wondering whether Eos ever actually did guard Albert's possessions, as she does in the Landseer portrait, or whether the picture's composition is purely symbolic. Another question that's crossed my mind is whether Albert struck his dog with the same cane Eos is carefully guarding; according to the biographer Jules Stewart, the prince "took a severe hand to his children's upbringing" and "could easily sink into ill-temper."

Grisby's never been known to guard any of my possessions, though since we go for a run together every week, he's grown especially attached to my running shoes (so much could be said about what shoes signify to dogs), not only because they're sweatier and smellier than most of my other shoes but also because they speak his language; he knows what they mean. Thomas Mann observed a similar skill in his dog Bashan: "He sees what my intentions are. My clothes betray these to him, the cane that I carry, also my attitude and expression, the cool and preoccupied look I give him, or the irritation and challenge in my eyes. He understands." Heatstroke is another anxiety when we go running since, like other flat-faced breeds, bulldogs are particularly susceptible to overheating and should be watched in hot weather (we never leave home without water, and I'm well versed in snout-to-mouth resuscitation). Still, he seems pretty resilient, and when he's well and truly beat, he'll flop down in the shade and refuse to move. Unlike many of us, he knows what he can take.

Our running loop in Baltimore is in a wooded area of Druid Hill Park behind the zoo, where, although we're right in the center of the city, we'll sometimes encounter foxes, deer, box turtles, and rat snakes. On our run, we follow a densely wooded path, which sometimes, after a storm, will get blocked by a fallen tree. When this happens, I climb over the trunk, and Grisby follows me, usually with a little trepidation, half jumping, half climbing, and landing on the ground each time with a satisfying snort.

With jogging as with walking, Grisby likes to keep his own

pace; sometimes he'll run ahead of me, sometimes he'll fall behind. He likes to investigate wayside smells, chase squirrels and
rabbits, take shortcuts, and greet strangers. On this particular
path, people generally have their dogs off leash, and we always
go early, when we're least likely to run into people who might
not appreciate a French bulldog's exuberant greeting. For this
reason, too, we stick to the more secluded areas of the park. The
northern end, which contains some of the oldest forest growth
in the state of Maryland, is a natural wooded habitat. Here, undergrowth covers a crumbling man-made pond, and the roads
are closed to traffic. Sometimes we meet stray dogs wandering
around in the woods; there's apparently a small feral population,
mostly pit bulls; they've all been friendly so far.

Our morning run, when we have one, is always the best part
of my day. Nothing ever goes wrong. Grisby's enthusiasm is never
dampened. Then I go home or to work, and life goes on in the
usual fashion, which is to say it's full of drawbacks, hesitations,
disappointments, arguments, and anxieties, the kinds of things
that never bother me when I'm running with Grisby. I love being
in the park with him; he loves being there with me. It's really as
easy as that.

FLUSH

"HE & I ARE inseparable companions," wrote Elizabeth Barrett of her cocker spaniel, Flush, "and I have vowed him my perpetual society in exchange for his devotion." The poet kept her promise and remained committed to her pampered spaniel until he died, at a healthy old age. Her loyalty was, she wrote, the very least she could do, since Flush had "given up the sunshine for her sake."

The young spaniel was originally a gift from Elizabeth Barrett's friend Mary Mitford in 1842, given partly to help ease the poet's grief after losing two of her brothers in one year and partly to relieve her loneliness, as she was bedridden with symptoms of consumption. When Flush arrived in her life, Barrett, aged thirty-five, was spending almost all her time in an upstairs room in her family's London home at 50 Wimpole Street; her delicate health meant she rarely saw anyone other than her immediate family and their household servants. The popular perception of Barrett before her marriage, like that of her contemporary Jane Welsh Carlyle (see NERO), is of a lonely, childless, unhappy middle-aged woman for whom her dog was a compensation and substitute for human love, yet the situation of both women was far more subtle and indeterminate than this easy cliché suggests.

Flush is best known to us not through Barrett's letters, in which he plays a major role, but through Virginia Woolf's *Flush: A Biography* (1933), created, in part, as a playful parody of the popular Victorian life histories written by her friend Lytton Strachey—books like *Eminent Victorians*, *Queen Victoria*, and *Elizabeth and Essex*. In this charming story, told from the spaniel's perspective, Woolf makes use whenever possible of Elizabeth Barrett's and Robert Browning's own words, drawn mainly from their letters. "This you'll call sentimental—perhaps—but then a dog somehow represents—no I can't think of the word—the private side of life—the play side," she wrote to a friend, which perhaps explains why she later dismissed *Flush* as "silly . . . a waste of time." Nevertheless, it remains one of her most popular books.

Part of the reason for its popularity, I'd suggest, is that *Flush* is really and truly about Flush, and not his human companions. Without wanting to generalize too much, I've noticed that dog books often have much more to say about humans than they do about dogs. In many cases, it seems, those who write about their dogs are actually writing about something else entirely—their families, their childhoods, or their bonds with nature. Or perhaps they're writing about dogs as a way to remind us to appreciate the simple things in life, to enjoy the kinship claim of animals, or to accept the latter half of life with grace and dignity. In such books, the dog's purpose is to catch the attention of the reader and, like Hitchcock's famous MacGuffin, to drive forward the human plot. "Much more than a dog story," reviewers will say, as if a dog story by itself is so very little.

In Woolf's version of the tale, after a playful puppyhood spent in the English countryside, Flush soon reconciles himself to a quiet life with his invalid mistress, waking her in the morning with his kisses, sharing her meals of chicken and rice pudding soaked in cream. Elizabeth's health improves when she meets Robert Browning, though Flush feels neglected at the intrusion, and can't restrain himself from biting Browning in a fit of jealousy when he calls round one afternoon to pay a visit. Afterward, according to Elizabeth, Flush "came up stairs with a good deal of shame in the bearing of his ears." His mistress refused to forgive him until eight o'clock in the evening, when, she writes to Browning, having "spoken to me (in the Flush language) & . . . examined your chair, he suddenly fell into a rapture and reminded me that the cakes you left, were on the table."

Incidentally, Flush has no reason to complain of being displaced by Mr. Browning, whom Elizabeth loves very differently from the way she loves Flush. In many ways, the dog always comes first in her affections, but her love for Flush is more protective and maternal than erotic. Elizabeth writes to Mary Mitford that unlike other dogs, Flush dislikes bones; prefers sponge cake, coffee, and partridge cut into small pieces fed to him with a fork; and will drink only from a china cup even though it makes him sneeze. His mistress sheds tears when Flush is kidnapped—as actually happened three times in real life (see BULL'S-EYE), though Woolf conflates the events into one incident—yet her grief is considered ridiculous ("I was accused so loudly of 'silliness & childishness' afterwards that I was glad to dry my eyes & forget my misfortunes by way of rescuing my reputation"). She finds it necessary, in a letter to Browning, to justify her tears: "After all it was excusable that I cried. Flushie is my friend—my companion—& loves me more than he loves the sunshine."

Men often seem to feel uncomfortable around "excessive" displays of emotion, especially those evoked by dogs. While he may not have ridiculed her tears, Robert Browning tried to persuade Elizabeth not to pay the ransom that was demanded for Flush when he was stolen. Elizabeth, in what was considered a reckless move by Browning and her family, went by carriage with Wilson, her maid, to the slums of Whitechapel to negotiate with the dog thieves. After five days and a payment of twenty pounds, much disapproved of by Barrett's father and her fiancé, Flush was returned to Wimpole Street. It seems interesting to note in this context that Virginia Woolf's husband, Leonard, appeared to share Robert Browning's attitude. Though he loved dogs, Leonard Woolf took a Cesar Millan–style approach to their training, intimidating them into "calm submission" before offering any sign of friendliness. His method was hardly a great success—the Woolfs' dog Hans was notorious for interrupting parties by getting sick on the rug, and Pinka, the dog Virginia Woolf used as the model for Flush, apparently ate a set of Leonard's proofs and urinated on the carpet eight times in a single day.

According to the social reformer Henry Mayhew, stealing dogs was in fact a commonplace racket in Victorian London. "They steal fancy dogs ladies are fond of," wrote Mayhew in 1861, "spaniels, poodles, and terriers." Jane Welsh Carlyle, the wife of the essayist and historian Thomas Carlyle, was also a victim of these cruel swindlers (see NERO). One day in June 1851, when Mrs. Carlyle was out walking with her husband and her Maltese dog Nero, "the poor little creature was snapt up by two men and run off with into space!" She doesn't want to give in to the gang, she writes, "for if they find I am ready to buy him back at any price (as I am) they will always be stealing him—till I have not a penny left!" Nero was stolen and returned three times, once managing heroically to escape and make his way home under his own steam. Later in the same letter, however, we discover why the dog thieves had such an easy time of it. Mrs. Carlyle complains that she might have to start keeping Nero on a chain when they leave the house ("and that is so sad a Life for the poor dog"), an observation that suggests he was seldom leashed. While there was far less traffic at that time than there is today, and while horse-drawn vehicles rarely reached the speed of automobiles, there were surely plenty of opportunities for dogs to get into trouble, quite apart from being snatched up by kidnappers. Leashes may be a drag, but city dogs aren't safe without them—in any century.

Barrett Browning never mentions any of her dreams, but since Flush slept in her bed (against doctor's orders), he no doubt played an active role in her dream life. Grisby appears in my dreams all the time, though not always in his usual form (and in dreams, as in life, he has to be taken out to answer the call of nature). Once I dreamed there were two Grisbys, identical twins who sat under my chair like little Cerberuses. In another dream, a larger dog followed him around everywhere. When I asked this dog's owner what his breed was, she replied, "He's a shadowboxer." This nocturnal doubling and shape-shifting seem unsurprising; as Freud himself remarks, in our dreams, "we are not in the least surprised when a dog quotes a line of poetry," though when we wake, we can't help trying to make sense of these strange transfigurations.

Most of my Grisby dreams, in fact, are nightmares—the expression, I assume, of all my repressed anxieties. They're always the same, and they return me to a primitive, prelinguistic level of distress—the kind of primal pain experienced by the child taken from its mother, or by the mother who loses her child. In my nightmares, Grisby is missing. I'm devastated, torn between going in search of him and getting a new dog right away, to cushion the pain. Sometimes I do one, sometimes the other, but whenever I get a new dog, it's always another male French bulldog, and I name him Grisby, too. Time passes. I grow to love my new

Grisby. The old Grisby is forgotten. Then comes the moment of horror: All of a sudden, I realize the original Grisby's still out there somewhere, all alone, lost, trying to get back to me. How could I have abandoned him? I'll wake in a sweat, and it always takes me a moment to realize that Grisby is right there in bed with me—the original Grisby, whimpering in his sleep. Does he dream of losing me? Has he moved on to a new Mikita, leaving me lost, dogless, alone?

Flush, too, moves on. By the time the Brownings have settled in Florence, he's grown accustomed to his new master, Robert, successfully making the transition from lapdog to family dog— something Grisby has never been able to do. Though he's lived with David all his life, Grisby is, categorically, my pet. "Your French bulldog," according to the author of *How to Raise and Train a French Bulldog*, "will bond with *one* member of the family," a line that David often repeats in a slightly affronted tone.

In the wider culture, this exclusivity is largely considered unwholesome; I've heard Grisby called a "mommy's boy," and when I wear a long skirt, he sometimes likes to hide under it. It's far healthier, according to popular opinion, for a dog to be part of the family. Family dogs are regarded as genial and good-natured, keeping guard over hearth and home. Free from pampering and protection, they romp with the kids each morning, nap in the sun all afternoon, then fetch Dad's slippers when he gets home from work. Unsurprisingly, family dogs appear most often in children's books, in which they love everyone unstintingly, demonstrating their loyalty by dragging old folks from burning buildings and saving kids from floods. Dogs like Lassie and Old Yeller spend their lives teaching families wonderful, life-enhancing lessons, and then, when they're no longer needed, go gently to the grave.

Unlike the snippy, jealous lapdog, the family dog loves everyone, regardless of age. In J. M. Barrie's play *Peter Pan*, Nana, a kind and docile black-and-white Newfoundland, keeps a close eye on the three Darling children, whose parents can't afford a real nanny. When the children are flying away, Nana howls to alert their parents, but her warnings are ignored, leaving Mr. Darling so remorseful that he sleeps in the kennel himself, until their safe return. Nana, usually played by an actor in a dog suit, was based on Barrie's own dog Luath, also a black-and-white Newfoundland, though, unlike Nana, Luath was a male. The author claimed he wrote *Peter Pan* "with that great dog waiting for me to stop, not complaining, for he knew it was thus we made our living." Still, when Luath discovered he'd been given a sex change in the play, wrote Barrie, he couldn't help expressing his feeling with "a look."

Other family dogs care for the elderly. In John Galsworthy's *The Forsyte Saga*, the dog Balthasar, a "friendly and cynical mongrel," develops a close relationship with Old Jolyon, the family patriarch. Originally one of a litter of three puppies, all of which looked so wrinkled and old they were named after the three wise men, Balthasar is part Russian poodle and part fox terrier, "trying to be a Pomeranian." The dog is found sitting protectively at Jolyon's feet when the old man dies in his sleep one summer afternoon under a tree. In his own memoir, written in later life, Galsworthy meditates on the importance of dogs in his life, speculating that "it is by muteness that a dog becomes for one so utterly beyond value. . . . When he just sits, loving and being loved, those are the moments that I think are precious to a dog."

According to Colette Audry (see DOUCHKA), certain breeds of dog are especially suited to this role. "Family men prefer poodles or cocker spaniels," she writes, "harmless creatures chosen specially to amuse the children, 'give them something to play with.' " As a feminist, Audry criticizes the way family patriarchs often reduce their underlings to the status of dogs. "The servant maintained a doglike silence, and the children romped about as though they were puppies," she writes. "Inevitably, one's love for one's wife became confused, up to a point, with the feeling one had toward a favorite dog or horse." Such a patriarch is the eccentric family farmer Dandie Dinmont in Walter Scott's novel *Guy Mannering*. Dinmont owns six long-haired terriers (along with "twa couple of slow-hunds, five grews, and a wheen other dogs") named "auld Pepper and auld Mustard, and young Pepper and young Mustard and little Pepper and little Mustard." When asked whether this is not rather "a limited variety of names," the farmer replies, "O, that's a fancy of my ain to mark the breed, sir." His fancy came true; this particular kind of long-haired terrier is now known as the Dandie Dinmont—the only example, to date, of a dog breed named after a literary character.

In Dickens's *Dombey and Son*, the loving dog Diogenes, largely democratic in his affections, is passed among the Dombey family. Originally owned by the schoolmaster Dr. Blimber, Diogenes is taken up by Paul Dombey, then given to his sister, Florence, after Paul's death, even though he is "not a lady's dog, you know," as Florence's admirer Mr. Toots explains, euphemistically. Diogenes is "a blundering, ill-favored, clumsy, bullet-headed dog, continually acting on a wrong idea that there was an enemy in the neighborhood, whom it was meritorious to bark at," and in order to get him into the cab to deliver him to Florence, Mr. Toots has to pretend there are rats in the straw. As soon as he's released from the vehicle, Diogenes dives under the furniture in Florence's house, dragging his long iron chain around the legs of chairs and tables, and almost garroting himself in the process.

Like many dogs in literature, Diogenes serves to indicate the character of the various people he encounters. Affectionate to Florence, Mr. Toots, Captain Cuttle, and Susan Nipper, he dislikes the sour Mrs. Pipchin and howls in her presence. Elsewhere the treatment of literary dogs can foreshadow human conduct. In Joyce Cary's short story "Growing Up," for example, the family patriarch returns from a business trip to find his young daughters appear unfamiliar and estranged. They express their violence first by mistreating Snort, the family dog, and then by turning on their father with homicidal aggression. This is the problem with being the household pet. If you belong to everybody, you belong to nobody, and you're surely better off as a lapdog than a scapegoat, however undignified you might feel.

GIALLO

IT'S REMOTELY POSSIBLE that, during her final years in Florence, Elizabeth Barrett Browning's dog Flush might have encountered Giallo, the Pomeranian belonging to the English poet Walter Savage Landor, the Brownings' friend and neighbor. In his younger days, Landor had lived in Florence for many years with his wife and children; he returned to England in middle age. Then, in 1858, as an old man of eighty-three, the poet found himself accused of libel, and escaped to Florence to avoid the resulting scandal. Here, the Brownings helped find accommodation close to their own home, the Casa Guidi, for this elderly gentleman and his frisky little dog.

In 1844, when he was living in Bath, Landor had a Pomeranian named Pomero, with fluffy white fur, bright eyes, and a yellow tail (Landor was said to be the model for Mr. Boythorn in

Bleak House, with Pomero transformed into a canary). A friend who knew Landor said he concentrated on the dog "all the playful affectionateness that made up so large a portion of his character. He loved that noisy little beast like a child, and would talk nonsense to him as to a child." "Not for a million of money would I sell him," wrote the poet. "A million would not make me at all happier, and the loss of Pomero would make me miserable for life." This loss came all too soon. "Seven years we lived together, in more than amity," mourned Landor, after his pet's death. "He loved me to his heart and what a heart it was! Mine beats audibly while I write about him."

When the poet returned to Florence in later life, his friend the sculptor William Wetmore Story gave him another Pomeranian, and this affectionate creature, named Giallo after his yellow fur, became Landor's closest companion for his remaining six years of life, causing him to be known by the locals as "*il vecchio con quel bel canino*" ("the old man with the beautiful dog"). Pomeranians are known for their loyalty and playful natures, but they also have less familiar advantages. In her 1891 essay "Dogs and Their Affections," the English novelist Ouida wrote, "The Pomeranian is a most charming small dog, and . . . there is an electric quality in his hair which repels dust and dirt."

In his lodgings on the Via Nunziatina, Landor soon became well known to English and American visitors, who described how Giallo's white nose would push through the door ahead of his eccentric master, and how Landor would take a seat in his armchair and hold forth on politics and literature, attributing his most controversial opinions to his dog. "A better critic than Giallo is not to be found in all Italy, though I say it who shouldn't," he claimed. "An approving wag of his tail is worth all the praise of all the Quarterlies published in the United Kingdom."

This popular and intelligent dog inspired many verses, including a couplet written on the occasion when Landor reached out and discovered his dog's nose was hot ("He is foolish who supposes/Dogs are ill that have hot noses"). Giallo was also the subject of the touching poem "To My Dog," written in August 1860, in which Landor acknowledged the fact that he would be in the grave long before his companion ("Giallo! I shall not see thee dead/Nor raise a stone above thy head"). He was right: Landor died in 1864, and Giallo lived on for another eight years in the care of his friend Contessa Baldelli. ("Poor dog! I miss his tender faithfulness," wrote the contessa when Giallo finally died in 1872.)

Landor obviously enjoyed using his dog as a mouthpiece for controversial political and literary opinions. Since the poet's sentiments were largely unpopular—he had a patrician contempt for the masses, for one thing—it was an inspired strategy to attribute his own words to a fluffy Pomeranian. As Virginia Woolf no doubt discovered while writing *Flush*, there's a special pleasure to be had in putting words into a dog's mouth—and many others have done so, though rarely with Woolf's eloquence.

For the most part, canine correspondences are private games between friends or family members. On December 3, 1855, Jane Carlyle, the wife of Thomas Carlyle, noted in her diary that she wrote "a pretty long letter" from her dog Nero to her friend Mrs. Twisleton ("Oh Madam; unless I *open my heart* to someone; I shall go mad—and *bite*!"). The dogs in Sigmund Freud's family "wrote" birthday poems to their master every year, rhymes that were actually composed by Freud's daughter Anna. According to the author Roger Grenier, Marcel Proust wrote regular letters to Zadig, the dog that belonged to his lover Reynaldo Hahn ("My dear Zadig, I love you very much because you are soooo sad and full of love just like me"). It's not difficult to interpret these communications as ways of expressing separation anxiety and other infantile feelings that are normally suppressed.

Things get more complicated, however, when people speak on their dogs' behalf in a more public way, especially online. These days, there are thousands of journals on the Internet purportedly written by dogs, many linked to dog-themed social networking sites like Dogster or Dogbook (if Walter Savage Landor were alive today, he might well have blogged as Giallo). Their voices are uncannily similar: playful, enthusiastic, and endearingly dim-witted—the voice of a loving but backward child. This is also the personality attributed to Clark Griswold, the German shepherd star of a viral YouTube video in which the dog is teased playfully by his owner about not getting his favorite treats. Clark's reactions, dubbed by his owner, are absurdly disconsolate. His engaging credulity is shared by the unnamed bulldog of the website Text from Dog, whose messages to his owner swing from engagingly exuberant ("I have to tell you . . . You accidentally fed me TWICE this morning . . . I GOT TWO BREAKFASTS . . . THIS IS THE GREATEST DAY OF MY ENTIRE LIFE"), to coyly passive-aggressive ("Who's MORE important: ME or your girlfriend?"). There seems to be some kind of unspoken agreement that, in terms of personality, dogs are adorably dense (as opposed, perhaps, to smart, snooty cats).

According to Stanley Coren, a dog behaviorist and psychologist at the University of British Columbia, dog blogging is "a sign of affection," and "trying to adopt a dog's point of view can be a healthy exercise" for pet owners. "If we love them dearly, we're always trying to crawl inside their heads and figure out what's going on," suggests Coren. "And if we love them dearly enough, we want other people to share in the dog's expertise." This, to me, seems an oddly disingenuous response to a phenomenon that begs for deeper consideration. Surely it's obvious—is it not?— that these dog blogs do not actually "adopt a dog's point of view" or "share a dog's expertise." The fact is, dogs have very little to do with them. These daily chronicles, with their infantilized voices, present-tense observations, and phonetic spellings, are produced by and for adult *Homo sapiens*. There are no pets online, just projections and displacements, human fantasies, and a willful return to the affections and appetites of childhood.

In *Pack of Two*, Caroline Knapp quotes Susan Cohen, the director of counseling at New York's Animal Medical Center, who is fascinated by the way people talk about (and on behalf of) their dogs. "When someone offers what sounds like a human interpretation of a dog's behavior," says Cohen, "it gives you something to explore. It might not tell you a lot about the dog, but it helps tell you what the person is thinking, what they're hoping, fearing, or feeling." When the dog is owned by a couple, for example, its voice can be used by the "parents" to accuse each other of neglect ("Mommy found my long-lost tennis ball—you know, the one that Daddy lost and didn't bother to replace") or to portray themselves as unconditionally lovable ("Oh, Daddy's dirty socks smell so good!"). At the most basic level, the "dog" here may be the blogger's infant self, beloved by Mother without reservation, no matter how odd he or she might look or smell. In the purported form of a dog, the infant self can express needs and feelings that the adult ego might, with good reason, want to distance itself from. On the Internet, no one knows you're *not* a dog.

The same dynamic also applies to in-person interactions. Arnold Arluke and Clinton R. Sanders note how pet owners, when "deciphering" their animal's symptoms for veterinarians, will "explain" their companion's moods ("She's upset that we have a new baby"), speak dyadically ("We aren't feeling well today"), or speak for the animal itself ("Oh, Doctor, are you going to give me a shot?"). Whether online or face-to-face, to speak in the voice of your dog is to engage in an act of self-deceiving ventriloquism, allowing you to be at the same time both beloved child and adoring parent. In this voice, you can buffer complaints, elicit apologies, confess wrongdoings, and mediate outlawed or forbidden impulses.

I can't help noticing that I'm writing about "your dog," analyzing what "people" do. Writing in the second or third person is another way to create distance from things that feel uncomfortable for us—that is, for me—to confess. The truth is, I'm so used to articulating Grisby's preferences that it's difficult to admit they're not far from my own: He loves coffee cake but dislikes asparagus, likes cartoons but gets bored by foreign films, likes white bread but not tortillas. Moreover, last Christmas, writing in my left hand, I added "Grisby's" shaky signature to mine and David's at the bottom of our Christmas cards, reversing the *R* as if he were still learning to write. I've projected onto Grisby, it seems, the stereotype of a child with Down syndrome: comical, captivating, and always up for a cuddle.

Such children are famously lovable, but adults with the same condition are often shunned, especially since weight gain is a common side effect of neurological medication. Similarly, precocious children can be delightful, but infantile adults are disturbing, their sexual maturity sitting uncomfortably beside the child's lack of self-restraint. In the same way, dog owners who write or speak as their dogs can do so comfortably only when their dogs are "fixed"; a blog or video giving human voice to a dog's sexuality would be not cute but unsettling. Chop off his balls, however, and he can be a fat child forever.

I can't speak on behalf of other dog owners (or their dogs), but I suspect I've given Grisby this kind of personality as a way of connecting my adult and childhood selves. He is, in other words, a "transitional object"—a phrase coined by the child psychoanalyst D. W. Winnicott to mean a personal possession, like a teddy bear or security blanket, that helps the child feel safe away from home. Transitional objects are not limited to childhood. Adults, too, need things that remind them of their private worlds: personal photographs used as screen savers, lucky charms, religious icons, sports mascots—anything with a stable meaning that can avert loneliness, mediating between the familiar world of home and the impersonal workplace or public realm.

Dogs make very handy transitional objects because we can use them as outlets for all kinds of different emotions. In my case, Grisby forms a bridge between my inner life and the "real world" out there, toward which I'm increasingly ambivalent. On the one hand, I want to function successfully as an adult in the wider world; on the other hand, I want to stay at home, regress to infancy, and keep the outside world at bay. It's always easier to make this difficult transition with a friendly bulldog by my side.

HACHIKŌ

HIDESABURŌ UENO, A professor of agricultural science at Tokyo Imperial University, always took the four o'clock train home from work, and every day, his dog, Hachikō, would be waiting for him on the platform at Shibuya Station. When Professor Ueno died of a cerebral hemorrhage in May 1925, in the middle of a lecture, his gardener, who inherited his house in the Kobayashi district, also adopted Hachikō, and for the next ten years, this golden-brown Akita would return to Shibuya Station every day to meet the four o'clock train, hoping to see his beloved master again. In 1935, Hachikō's body was found in a Tokyo street. His remains were stuffed, mounted, and put on display in Japan's National Science Museum. A bronze statue of the famous dog stands outside Shibuya Station to this day, and he also has his own memorial by the side of his master's grave in Aoyama cemetery.

Interestingly, Hachikō isn't the only dog whose statue oversees a railway station. In 2007, a memorial was erected at the Mendeleyevskaya station on the Moscow Metro in honor of Malchik, a stray mutt who lived there for about three years, becoming popular with commuters and Metro workers. Malchik claimed the station as his territory, protecting travelers from drunks, the homeless, and other stray dogs. In 2001, after getting into an altercation with a bull terrier, Malchik was stabbed to death by the other dog's owner, a psychiatric patient with a long history of cruelty to animals.

Heroic dogs like Malchik and Hachikō continue to appeal, growing even more famous and beloved as time goes by. Hachikō has been the subject of two children's books and two movies, the more recent of which—*Hachi: A Dog's Tale*—starred Richard Gere as the Professor Ueno figure. In 1994, millions of radio listeners tuned in to Nippon Cultural Broadcasting to hear a newly restored recording of Hachikō's bark, and in 2012, enthusiasts queued for hours to see an exhibition of rare photographs from the dog's life. To his fans, Hachikō was unique, perhaps even miraculous in his devotion. Yet to those with a broader view, Hachikō's story is simply the most recent variant of an ancient and widespread folktale motif: the Faithful Hound.

Arguably, the best-known example of this theme may be found in the famous Welsh story of the hero Llewellyn and his loyal hound, Gelert. In this legend, Llewellyn returns from hunting to discover his baby missing, the cradle overturned, and blood around the mouth of his dog. In a rage, the hero draws his sword and stabs Gelert. The dog's expiring whimper wakes the baby, who's lying unharmed under the cradle along with a dead wolf, killed by the loyal Gelert. Overcome with remorse, Llewellyn buries the dog with great ceremony, yet the sound of his faithful dog's dying whimper haunts the impulsive hero from that day forth.

A tale that remains current in so many different times and

cultures must speak to a universal truth. Dogs are certainly faithful creatures, perfectly capable of expressing grief for a lost companion, spending time beside the dead body of a beloved friend, and finding their way home despite tremendous obstacles. Nevertheless, the primary instinct of a healthy dog—like that of all healthy creatures, including humans—is self-preservation, and on further investigation, these miraculous hounds often prove rather less marvelous than they first appear.

The nineteenth-century counterpart of Hachikō was Greyfriars Bobby, a humble little Scottish terrier who lived in the Greyfriars burial ground in Edinburgh, allegedly unwilling to leave his master's grave. In his book *Greyfriars Bobby: The Most Faithful Dog in the World*, author Jan Bondeson proves there were actually two different dogs going under Bobby's name, both ordinary mongrels who made their home in the burial ground, eating leftovers provided by a local restaurateur. Bobby, explains Bondeson, was simply the latest in a series of now-forgotten cemetery dogs, including Médor, Dog of the Louvre; the Dog of Montparnasse; and the Dog of the Innocents. Like Greyfriars Bobby, these turned out on closer inspection to be "composite" dogs, strays that unconsciously took advantage of the public's willingness to bring them scraps, helping—as they believed—to sustain these loyal creatures in their lonely vigils.

In the case of Hachikō, the dog seems to have been feeding—quite literally—on his own publicity. Once he became a regular visitor to Shibuya Station, people started to spread the word. The dog's story was reported in the local paper, and commuters who passed through Shibuya began to look forward to seeing Hachikō, saving him leftovers from lunch. Since he spent less than two years meeting Professor Ueno at Shibuya and another eight years repeating the behavior, it seems fair to suggest the dog's daily trip to the station may have been motivated less by steadfastness than by the anticipation of regular snacks.

The universal truth to which all these stories speak is not that of the dog's miraculous fidelity but rather our need to believe in it. Dogs like Hachikō are symbols of canine commitment, affirming our confidence not only that dogs are capable of intense devotion but also that we are worthy of inspiring it. Yet while there's something marvelous about the apparent dedication of these long-suffering dogs, it can also seem unbearably poignant. I was a child of seven or eight when I first heard the tale of Llewellyn and Gelert from my grandfather, and it upset me so much I had to make a conscious effort to forget it; I didn't want to carry around such a terrible story in my head. When I think about the tale now, however, my anxiety takes refuge in its cracks and flaws. What kind of wolf is small enough to be hidden by an overturned cradle? Wouldn't Llewellyn look for his missing baby before jumping to conclusions? And where was Mrs. Llewellyn?

Some people feel uncomfortable with the emotional burden placed on them by devotion of such magnitude. The poet Rainer Maria Rilke loved animals but avoided dogs because, he claimed, they required too much from him. Their utter dependence on human beings was too painful for him to bear. "Why did dogs make one want to cry? There was something so quiet and hopeless about their sympathy," wonders the protagonist of Daphne du Maurier's novel *Rebecca*, when she observes Jasper, one of her husband's spaniels, "knowing something was wrong, as dogs always do." Such moments bring us so readily to tears, I suggest, because they're evocative of primal emotion. They provide a screen onto which we can't help projecting memories of a time in our lives when we ourselves were most doglike.

As babies, we, too, were once full of inexpressible emotion and inarticulate yearnings. In *Pack of Two*, author Caroline Knapp argues that, with our dogs, we develop a bond similar to the magical, once-in-a-lifetime connection we once experienced with our mothers. "I sit outside with Lucille day after day," writes Knapp, "and I feel that torrent of emotion—joy and delight and surprise along with self-doubt and anxiety and confusion—and I think: This is love, pure but not simple." Around Lucille, Knapp can't help indulging in projection. As she explains, she unconsciously ascribes to Lucille her own thoughts, feelings, and desires, often those she finds difficult to acknowledge.

Pets—just like dolls, cuddly toys, babies, and small children—are especially suitable for projection since they're unable to speak or let us know their feelings in any detailed way, which is why they tend to arouse our fierce love and protective instincts. Bulldogs and other flat-faced dogs may invite more projection than other breeds because their faces are so human, with the eyes at the front of the head rather than the sides. This gives them another advantage, too. According to the anthrozoologist Hal Herzog, "breeds with short snouts like bulldogs, boxers and pugs understand human signals better than long-nosed breeds like Dobermans, dachshunds, and greyhounds," owing to the frontal location of their eyes.

The projection of human emotions onto animals, rationalists would argue, is a kind of anthropomorphism, a cardinal sin among those who study animal behavior. Traditionally, scientists have criticized anyone who ascribes to animals "human" characteristics such as consciousness, desire, intelligence, goals, emotions, and preferences. Some go so far as to put words like "pain" and "hunger" in quotation marks, as though animals were incapable of feeling even these basic sensations. Outside the scientific domain, too, we sometimes use special terms when animals, not humans, are the subject—not "kill" but "cull," not "babies" but "litter," not "chop off" but "dock," not "lovemaking" but "mating." This makes me think of the philosopher René Descartes's claim that animals are like machines, without feelings or emotions, and the sounds they make when "in pain" are simply the noises of a broken object, meaningless and automatic. While we may not know exactly what animals are feeling, and while it may not be wholly accurate to use the same terms of animals that we use of ourselves, in the absence of any knowledge to the contrary, surely it's only right to assume they're as sensitive as we are to pain, loss, isolation, and abandonment.

Perhaps, in the end, those most primal feelings are all I can really assume about Grisby. Maybe everything else I see in him—what I normally think of as his "personality," his quirky charms, curious tastes, and funny habits—is projection. I claim I love him for his own special qualities, his unique Grisbiness, yet if I had to be more specific, all the characteristics I could mention would be human concepts: sweetness, playfulness, good humor, charm. This makes me wonder whether what I love about Grisby is the way he seems, as so many dog owners say of their pets, "almost human." More to the point perhaps, is he really so sweet and loyal? Is he, in fact, eager to please me, or is he more interested, like the rest of us, in pleasing himself? If acting in a way that makes me happy means something good will happen to him—that he'll get a treat, some crumbs, a kiss, a scratch behind the ears—then he'll naturally be more likely to do it. I wonder: Is human "altruism" any different?

Curiously, from time to time, I'll get a glimpse behind the veil of my own projection and see Grisby "as he really is"— a slightly overweight eight-year-old French bulldog, no more nor less. It happened once in training class, when from the sidelines, I observed him in a lineup of other dogs, and for a moment he was just a dog. Then, suddenly, he was *my* Grisby again. It was like looking at one of those optical illusions that seem to switch in your brain—you see a rabbit and not a duck; then suddenly all you can see is a duck, however hard you try.

This happens most clearly in the case of photographs since the camera records impressions with impersonal objectivity. I have one or two snapshots that capture something fleeting of the

Grisby I know, but he really doesn't photograph well. More accurately, he doesn't photograph at all. Trying to capture *my* Grisby in a picture is like trying to photograph a ghost. What appears in the frame is merely the image of a dog—not *my* Grisby but the dog-in-the-world, the everyday animal. I'm also made aware of my projection whenever I get back from a trip and the creature that greets me isn't *my* Grisby at all but just a brown-and-white French bulldog. For a moment, it's as though I'm seeing him naked, exposed; then all of a sudden, the veil of projection falls, and all at once he's *my* Grisby again. Colette Audry describes a similar experience when she rejoins her German shepherd Douchka after weeks apart. The dog, for a moment, seems less herself. "It struck me afterward that my pleasure at seeing her again had not been so great as I'd anticipated," Audry observes. "Was that what she really looked like? My memory must have embellished her in retrospect."

As psychologists have often testified, our feelings for our dogs can get tangled up with all kinds of issues and complexes involving residual jealousy, childhood traumas, the need for attention and affection, and feelings of rejection and despair. Caroline Knapp points out that "a person's core sense of self—anxieties, insecurities, grandiosity, fantasies of self and other—can come bubbling up when it comes to controlling a dog, even in the smallest and seemingly most inconsequential ways." For Knapp, these anxieties clustered around the question of her authority, which, she felt, was regularly challenged by her dog Lucille. "When the dog fails to come on command, when she ignores me," observes Knapp, "I feel some well of fear rise up about being inadequate, unworthy of attention, out of control. I can hear a small voice inside: She doesn't come when you call, because she knows you're a wimp. She doesn't come because she doesn't love you. You're a loser; you can't even control your dog."

My own particular anxieties seem to cluster around the issue of abandonment and neglect. I hate leaving Grisby at home. I especially hate leaving him alone; I take him almost everywhere I go, even to places where his presence is no doubt inconvenient for others: pubs and cafés, readings, lectures, movies, faculty meetings, cocktail parties. (Right now, as I type, he's sitting beside me in a booth at a dog-friendly coffee shop.) Whenever I have to leave home without him, I get a horrible sinking feeling in the pit of my stomach. I imagine him sitting just inside the front door where I left him, waiting for me to return. I think of him wondering where I am, feeling alone and unloved. I know these anxieties are mine, not his. I know that if I disappeared tomorrow, Grisby would eventually get used to being with someone else. He might even be better off with someone who gives off fewer nervous vibrations. Perhaps this is what I'm truly afraid to face: that as long as he's got food and company, he might barely notice I've gone.

ISSA

ISSA, WHOSE NAME translates from the Latin as "her little la-dyship," was a small white dog, widely believed to be a Maltese, immortalized in descriptive verse by the Roman poet Martial. In his poem, Martial makes light fun of the bond between Issa and her master, a well-known Roman figure named Publius, who hasn't been conclusively identified by historians but is generally thought to be the Roman governor Publius of Malta (who, since he lived in Malta, may well have owned a Maltese).

"Publius' darling puppy," writes Martial, is a "modest and chaste little lap dog, more coaxing than any maid." She's discreet and genteel, "purer than a Dove's kiss," thoroughly fastidious, and perfectly toilet trained: "When overcome by nature's long-ing, never by one drop does she betray the coverlet." To her mas-ter, she's worth "all the costly pearls of India." He's even had her

portrait painted, Martial informs us, so "death should not rob
him of her altogether."

Classical scholars have mixed opinions about this epigram-
matic poem. Some consider it to be a fond tribute to Publius and
Issa; others see it as mocking and cruel. Either way, Martial's
allusions suggest the Roman's affection for his cosseted chum
was common knowledge at the time. Adding to the incongruity,
the Maltese, with its silky white fur and fastidious temperament,
was known in classical times as the Roman ladies' dog and seen,
like most lapdogs, as a useless luxury, often a symbol of women
themselves, with their misplaced values and susceptibility to the
vagaries of fashion. Dogs like Chihuahuas, Yorkies, and shih tzus
have always been regarded as typical women's pets, as opposed
to "men's dogs" like the macho Doberman, butch Rottweiler,
and virile mastiff. The Maltese is "admirable, beautiful," writes
Ouida in "Dogs and Their Affections," "and his aristocratic appearance, his little face which has a look of Gainsborough's and
Reynolds's children, his white silken coat, and his descent from
the darlings of Versailles and Whitehall, all make him an ideal
dog for women."

If the Maltese is feminine and the bulldog male, the French
bulldog is a confusing combination. At a solid thirty-two pounds,
Grisby is a muscular minion, both lapdog and chaperone. An article in the *New York Times* style section from 2005 describes the
breed as "gay vague" (as opposed to the Boston terrier, which signifies "straight," and the Jack Russell, which is apparently out-
and-out "gay"). Greenwich Village, where I first noticed them,
has a high population of French bulldogs, and it's true that, as
small apartment dogs that are also solid and butch, they have a
particular appeal to gay men. If they were "gay vague" in 2005,
moreover, a brief trawl online suggests that in the last nine years,
they've swung further to the far end of the spectrum: comment-
ers on various dog-themed discussion boards refer to French bull-

dogs as "übergay," "a gay stereotype," "a gay guy's doggie," and "the gayest dog ever.".

Would we vote for a president who owned a French bulldog? While it's certainly unusual for a leader to be known for his devotion to a lapdog, owning a pooch can make even the most intimidating public official seem more human, as public relations advisers are well aware. It's apparently still the case that a president's popularity increases whenever he's photographed with his dog—as long as he's not picking it up by the ears, the way Lyndon Johnson notoriously hoisted his unfortunate beagle. For this reason, the First Dog has always been as much a fixture of the White House as the First Lady. In his 1952 "Checkers speech," Nixon, then a candidate for vice president, won the public's confidence by admitting he'd broken the rules and accepted one single personal gift from a political donor: an American cocker spaniel named Checkers. "I just want to say this, right now, that regardless of what they say about it, we are going to keep it," he asserted, swaying public opinion determinedly in his favor.

The Checkers speech was, predictably, compared to the "Fala speech," a public address made by Franklin D. Roosevelt in 1944. President Roosevelt was running for his fourth term when rumors surfaced that he'd accidentally left behind his Scottish terrier, Fala, when visiting the Aleutian Islands—and had sent back a navy destroyer to pick him up, at the taxpayers' expense. Roosevelt made fun of these rumors in a speech that allegedly helped secure his reelection. "You can criticize me, my wife, and my family, but you can't criticize my little dog," complained the president. "He's Scotch, and all these allegations about spending all this money have just made his little soul furious."

There's something about people's relationships with their dogs that seems essentially honest, which is why it always boosts our confidence to see the president interacting with his pooch. On one level, of course, such appearances are simply excellent photo ops; on another level, however, the president's relationship with his canine pal seems much more personal than, say, his relationship with his wife or children. The First Lady, like her husband, is a public figure, part of the trappings of office; the president's offspring, too, are encouraged to show their support before the cameras. Motiveless alliances are unknown in politics. Only the president's dog exhibits authentic and unaffected allegiance; only his dog doesn't know he's the president.

It's widely believed there's a clear link between cruelty to animals and violence toward human beings. This would suggest those who express affection for animals are peaceful people, which isn't always the case (look at Hitler and his dog Blondi). Still, when it comes to political leaders, dogs provide a fascinating axis between public and private lives, and there's something both reassuring and slightly perverse about pictures of Fidel Castro, Leon Trotsky, and Henry Kissinger petting their dogs. Most of the time, these are large, robust beasts, of the kind that make a striking picture romping on the lawns of mansions and palaces. Owing to his small stature, George W. Bush's Scottish terrier Barney drew unmerited contempt. When Bush introduced Barney to Vladimir Putin, the Russian president "kind of dissed him," according to Bush. Later, when Bush met Putin's black Labrador Koni, Putin made sure the US president noted that Koni was "bigger, stronger, tougher, and faster than Barney." One wonders what Putin would have made of Issa.

These days, First Dogs are given simple, democratic, and frankly unimaginative names (Buddy, Barney, Bo), often chosen by the president's kids or by public poll. This was not always the case; former canine residents of the White House have possessed names that today's public relations advisers would almost certainly consider out of bounds. George Washington had four black-and-tan coonhounds, named Drunkard, Taster, Tipler, and Tipsy; he also had American staghounds called Sweetlips and

Scentwell. Calvin Coolidge had a terrier named Peter Pan. Theodore Roosevelt had a Manchester terrier called Blackjack. John Adams had a mixed-breed named Satan.

The best presidential dog's name, in my opinion, is that given by James Garfield to his large black Newfoundland: Veto. Some said the dog was named in honor of President Rutherford B. Hayes, whom Garfield had helped to sustain a record number of vetoes (five bills in three months). Others saw the dog's name as a warning from Garfield to the rambunctious Congress of 1881, letting the legislators know that he was under no obligation to sign all the bills they were trying to pass. While Garfield no doubt enjoyed plenty of jokes about exercising his Veto, the president's assassination six months into his term meant he never got the chance to do so—at least not in the political sense.

A veto is the power to stop an action, which is something Grisby does all the time, though seldom without my consent. He prevents me from going away for long periods of time; he prevents me from taking spontaneous trips; he's the reason I no longer attend out-of-town conferences. He serves as a buffer or barrier, preventing me from getting too close to other people, keeping the world at bay. As he's always with me, anyone engaging with me also has to engage with him, at least on a superficial level—by asking his name, for example, and commenting (favorably, of course) on his appearance and behavior. My feelings for Grisby can sometimes get in the way of my feelings for other people in that, if a person interacts with me and ignores Grisby, I find it rude and off-putting. If I'm asked to leave him at home, I can't help feeling hurt. I'll often leave social functions early, using Grisby as an excuse (both when he's with me and when he's not). I realize some people probably find this behavior annoying, just as I find it irritating when anyone turns up with a young child in tow.

It also happens, on occasion, that Grisby becomes the target of hostility that, I can't help feeling, is actually—perhaps unconsciously—meant for me. For a while, he became the focus of a minor controversy at work involving the presence of pets at departmental meetings. On such occasions, the presence of a lunch buffet means that, according to the college's pet policy, the meeting is officially off-limits to dogs. Yet this rule was overlooked for many years; a colleague's popular and well-behaved pet, Gelbert, was in regular attendance. But Grisby is no Gelbert, and for some reason he—or whatever he seems to represent— seems to rub certain people the wrong way. It didn't help, I confess, that he once vomited spectacularly during a guest's presentation of the new online software, but dogs get upset stomachs just like the rest of us, and it hardly seems fair to hold them accountable. However, when a colleague publically expressed her fear of Grisby—even confessing that she kept a knife in her purse in case she should ever need to defend herself—I couldn't help wondering who (or what) was the real target of this irrational phobia. Surely it couldn't possibly be the affectionate little creature sleeping sweetly under my chair.

It's true that, when he was a puppy, Grisby, like all puppies, caused his share of trouble, but all damage was restricted to the home. The size of a small rabbit (not including his enormous ears, which he spent years growing into), he spent his first months of life destroying our possessions and ruining our rugs. Among the treasures he chewed to bits: a deck of rare tarot cards, my favorite pair of glasses, and a vintage volume of the Dos Passos *U.S.A.* trilogy, the thought of which still quickens David's pulse even today. We couldn't leave him alone until he was fully toilet trained, which took almost nine months. He could never manage a quick stroll around the block; he had to stop and investigate everything—every smell, every scrap of paper, every puddle, and every tree. To make matters worse, the ground floor of our building is used for wedding receptions on weekends, and I soon learned that a French bulldog puppy and a crowd of drunken, sentimental bridesmaids make a perfect storm. In the end I took to zipping him up inside my purse so I could sneak him out without his being spotted.

During her years spent as a radical politician, Colette Audry found her dog wasn't welcome at socialist resistance meetings. Although her artist friends were dog-friendly, her intellectual friends either didn't go in for dogs or, if they did, had "too great a human respect for the species to dream of bringing their own dog along." Audry responded by exercising her veto, dropping out of radical politics altogether to spend more time with her dog. When I worked in California and wasn't allowed to bring Grisby to work, I often felt like doing the same. Sometimes, after a particularly painful day, I'd stop at the beach on the way home, turn him loose, then stand back and watch him charge toward the ocean, causing mayhem on all sides. He'd knock over parasols, trample on sand castles, and kick dirt in people's picnics. After stifling my misery all day at work, I always got a thrill watching my avatar wreak joyful destruction on humanity.

Yet if Grisby acts as a buffer, he also acts as a bridge, keeping me connected to the outside world. He gives me a reason to do things I'd have no interest in doing on my own: running in the park, Rollerblading around the tennis courts, swimming at the lake. Sometimes, in the days before Grisby, I wouldn't leave the house for days on end—now I'm up early every day for his morning constitutional. He connects me with other people, serving as a talking point and an icebreaker. If he gives me the power of veto, he also forms plenty of democratic coalitions, and at the same time, paradoxically, treats me as though I rule the world. While it may very well be true that the president's dog doesn't know he's the president, on another level, it's a lie—every man's a president to his dog.

J I P

JIP (SHORT FOR GYPSY) is the spaniel belonging to David Copperfield's innocent and inept first wife, Dora Spenlow, in Charles
Dickens's 1850 novel, *David Copperfield*. Naughty and spoiled, Jip
behaves like most lapdogs in literature, challenging and menac-
ing his mistress's besotted lover, keeping his rival's passion at bay.
When David first approaches Jip, the spaniel "showed his whole
set of teeth, got under a chair expressly to snarl, and wouldn't
hear of the least familiarity." After their marriage, Dora and
David are almost penniless; nevertheless, Dora insists that "Jip
must have a mutton-chop every-day at twelve, or he'll die!" She
interposes the little dog between David and herself whenever her
husband wants to have a serious talk with her, and when David
buys his new bride a cookery book in the hope that she'll learn
some housekeeping skills, she uses it as a stool for Jip to stand on.

Jip signifies Dora's exasperating childishness. Untrained and entitled, this irritating little creature barks at tradesmen, chews geraniums, begs for toast, snaps, howls, hides from everyone except Dora, and strolls around on top of the dinner table during meals (see WESSEX). He embodies the chaos and disorder of the Copperfield residence, and her devotion to him characterizes Dora as a child-wife, incapable of handling adult responsibilities. Dora and Jip are closely identified as a pair of useless, pampered pets. After suffering a miscarriage, Dora grows sick and weak; Jip, too, quickly begins to fade. "He is, as it were, suddenly, grown very old," David observes. "He mopes, his sight is weak, and his limbs are feeble."

In strong contrast to the frail Dora Spenlow stands the independent, intellectual Dorothea Brooke, the heroine of George Eliot's novel *Middlemarch*. Wise beyond her years, the idealistic Dorothea has no time for lapdogs, as we discover when her suitor, Sir James Chettam, offers her a "little petitioner" in the form of a Maltese puppy. This breed, thought to be the oldest of the European toy dogs and established in England during the reign of Henry VIII, has always been popular among elegant women (see ISSA). Careful to differentiate herself from her more conventional sister, Celia, who "likes these small pets," Dorothea asserts to Sir James: "It is painful to me to see these creatures that are bred merely as pets. . . . I believe all the petting that is given them does not make them happy." Dorothea rejects the creature as she later rejects Sir James. She dislikes his fawning, just as she dislikes puppies' neediness. "They are too helpless: their lives are too frail," she complains. "A weasel or a mouse that gets its own living is more interesting."

"Ladies usually are fond of these Maltese dogs," muses the perplexed Sir James. He later gives the puppy to Dorothea's more appreciative sister, who eventually becomes his wife. Dorothea's indifference to lapdogs distinguishes her as intelligent and mature, unlike the frivolous Celia, whose love of puppies suggests that, at least in emotional terms, she's still a child. An article published in *Tait's Magazine* in 1856 reinforced the popular opinion that only "very young ladies" can keep pets without censure, as the animals may be needed as "diplomatic agents" during courtship. "Ladies of mature age" who insist on keeping a lapdog, continued the article, should be "brought to a sense of shame for the rather low level at which they have arrived."

In terms of cultural prejudice, we haven't come far. It's just as true today: in the arms of a fashionable young debutante, a well-dressed Chihuahua can be a charming fashion accessory, but in the lap of a lady in her sixties, it seems slightly sad. Women with lapdogs, according to popular opinion, are privileged, fussy, and indolent (as opposed to cat ladies, who are stereotypically disordered and unkempt). The lapdog is considered noisy and entitled, accustomed to riding in a designer carrier and eating gourmet food; it's taken for granted that the ladies who dote on lapdogs are sad spinsters or lonely aunts.

Lapdogs, in other words, are associated with unfulfilled maternal instincts. While most people accept dogs as part of the family unit, they often feel uncomfortable in the presence of a childless woman and a dog, as if only when a dog's not really "needed" can it be loved appropriately. By implication, then, lapdog-loving mothers are felt to have ambivalent relationships with their children, and in many cases this is true. The author Colette went nowhere without her French bulldog but rarely saw her only daughter, whom she left in the care of an English nanny. The childless Edith Wharton would keep her little dogs about her at all times, letting them join her at meals and drink from her teacup.

It's an unfair stereotype, of course—all kinds of people, including men, dote on their dogs—but rather than debunking it, my first impulse is to distance myself from it. I feel compelled to make it very clear that, although I love my dog to distraction, I'm not one of *those women*, and Grisby isn't one of *those dogs*. He's not a Chihuahua or a shih tzu; he's a tough little bulldog, too heavy to ride in a carrier or snuggle on my lap. I want to deny and disavow, to insist how different my situation is, instead of thinking about why it's so hard for a woman who buys sweaters for her dog to be taken seriously.

In some ways, after all, I *am* one of those women. I'm middle-aged, childless, and not officially married, and I dote on Grisby, who's a lapdog in all but size. He's certainly spoiled. I let him run off the leash, jump on the furniture, eat from my plate, sleep in my bed, and lick my face. I kiss him on the mouth and feed him anything he wants—hot dogs, pudding and cake, even ice cream. I've rocked him to sleep in my arms, thinking as I did so of the Duchess's baby in *Alice's Adventures in Wonderland*. ("If it had grown up," says Alice of the baby, "it would have made a dreadfully ugly child; but it makes rather a handsome pig, I think.")

I confess: I even buy him sweaters. This was something I resisted for years. After all, I told myself, he's all muscle—certainly strong enough to handle the cold. Then, one chilly winter's day a few years ago, I overheard a conversation at the grocery store checkout.

"Look at that poor little dog shivering in the cold like that," said the woman ahead of me.

"Mmm-hmmm," responded the checkout girl. "That dog need himself a sweater."

They were looking at Grisby tied to a parking meter outside in the snow. Noticing that he was, indeed, shivering, I realized he didn't look tough at all; if anything, he seemed pathetic. At that moment, I understood that my reluctance to buy him clothes had to do more with my comfort than his. Since then, he's accumulated a handsome selection of cold-weather wear, including a Burberry overcoat, a striped wool jacket, a plaid sweater, and a hooded yellow raincoat with little slits where his ears poke through. I love fastening the Velcro strap, pulling it tight around his bulldog belly.

It's wrong, I think, to assume that a dog is a substitute for a child, just as it's wrong to assume that dog love is necessarily of a different quality from human love. Still, it's often true that a first dog is purchased or given as a kind of compensation for a loss, as when a child is given a puppy to make up for the death of an earlier pet. In her short and charming book *Topsy*, originally published in 1940, Freud's friend and colleague Princess Marie Bonaparte describes being given her first dog as a way of making up for the loss of a beloved nurse. The princess refers to her first dog as a "talisman," explaining that a dog can often function as a protective substitute, sometimes replacing the mother, though it can represent anyone who was especially well loved.

People's love for dogs, then, may hint at a loss in their past, which is only one of the many ways in which canines can serve as clues. More obviously, a dog betrays its past treatment by its present behavior. By fawning, barking, or cringing, a dog can imply or accuse, inadvertently revealing the hidden dynamics of human relationships. It's no surprise that one of the archetypal functions of literary dogs is to act as a pointer, drawing the discerning readers' attention to signs they might otherwise overlook.

This motif finds its locus classicus in a Sherlock Holmes story— not the great detective's most famous canine mystery, *The Hound of the Baskervilles*, but "Silver Blaze," in which Holmes points out the "curious incident" of a dog failing to react to a mysterious visitor. When a guard dog doesn't bark at an intruder, Holmes remarks, it generally means he's someone the dog doesn't consider an intruder at all. The detective naturally respects canine skills in evidence gathering, occasionally borrowing his friend's dog Toby when he needs extra help sniffing down a clue. This hound may be "an ugly, long-haired, lop-eared creature, half spaniel and half lurcher," but Holmes defers to his expertise. "I would rather have Toby's help than that of the whole detective force of London," he comments in *The Sign of the Four.*

On other occasions, a dog will draw attention to a vital clue the human characters have failed to observe. In *David Copperfield*, Dora's secret engagement is accidentally revealed when the naughty Jip steals a love letter from her reticule, and her frantic attempts to get it back arouse her family's suspicion. By breaking all the rules, Jip is a living embodiment of Dora's unruly id. The dog enacts everything Dora strives to repress, sabotaging all her attempts at domestic routine, bringing chaos to the tea table. Elsewhere, it's not the dog itself that provides a clue but rather an item associated with it. A case in point is the ball left out by Bob, a wirehaired terrier in the Agatha Christie novel *Dumb Witness.* In this mystery, Emily Arundell, a wealthy spinster, is seriously injured when she trips over her pet's toy and falls downstairs—at least, this is what the evidence suggests. Yet as it turns out, Bob couldn't possibly have left his ball on the stairs, as he was outside all night; the "accident" turns out to be a crime perpetrated, it appears, by one of Emily's grasping young relatives.

Interestingly, the human transgressions inadvertently revealed by literary dogs are often sexual in nature, and usually committed by women: consider the folktale motif Dog Betrays Woman's Infidelity. An early version of this motif can be found in the anonymous thirteenth-century French romance *The Châtelaine of Vergi*, in which a lady accepts the love of an admiring knight on the condition that he never speak of their affair. To ensure secrecy, she instructs him to hide in the garden whenever he visits her, and to wait until her little dog comes out to fetch him. One day, during a public feast, she hears a joke about her "well-trained dog," and believes—wrongly—the knight has betrayed her, dying of despair. The motif is also found in La Fontaine's fable "The Wonder Dog," which is based on a story from Ariosto's fifteenth-century epic poem *Orlando Furioso.* In this tale, a fairy transforms himself into a magical lapdog in order to encourage a love affair between Argie, wife of a judge, and the chivalrous young Atis. The dog can perform so many amazing tricks that Argie decides she has to have the little creature at any price, even if it means betraying her husband (and it does).

In art as well as literature, lapdogs conceal or betray sexual secrets. In 1811, a portrait of Lady Caroline Lamb, who was the wife of the future prime minister William Lamb and would later have an affair with Lord Byron, was exhibited at the Royal Academy. At the time the portrait was being painted, Lady Caroline was involved with a young aristocrat named Sir Godfrey Webster, and the portrait, by Eliza H. Trotter, contains a coded reference to this affair. Lady Caroline is painted with her arm around a miniature bull terrier wearing two collars, one of which appears to be a bracelet of gemstones. Both the dog and the bracelet were gifts from Sir Godfrey; when her infidelity was discovered, she wrote, "I tore the bracelet off my arm & put it up with my chains in a Box. . . . I have written to desire some one will fetch the dog."

Anne de Cornault, the unhappily married heroine of Edith Wharton's ghost story "Kerfol," makes the mistake of giving her lapdog's collar to her lover, Lanrivain, as a memento. When her husband asks her about the missing collar, Anne says the dog must have lost it in the undergrowth of the park. Her husband says nothing, but that night, when Anne goes to bed, she finds something horrible: the body of her lapdog is lying on her pillow, "still warm." She takes a closer look, and "her distress turned to horror when she discovered that it had been killed by twisting twice round its throat the necklet she had given to Lanrivain." After this, every dog she takes for a pet ends up strangled on her pillow; in the end she "dared not make a pet of any other dog, and her loneliness became almost unendurable." Many years later, Anne de Cornault's estate, Kerfol, remains devoid of human presence, haunted by a pack of eerily silent hounds.

In Luigi Pirandello's play *Each in His Own Way*, the lusty Donna Livia owns a lapdog that expresses equal affection to two different men, revealing them both to be her secret lovers. In *The Great Gatsby*, George Wilson discovers—after her death—that his wife, Myrtle, had recently purchased "a small, expensive dog-leash, made of leather and braided silver"; he realizes she must have been having an affair since the Wilsons have never owned a dog. In Colette's short story "The Bitch," a soldier on leave in Paris calls unexpectedly on his mistress. She's not home, but the sheepdog he left in her care greets him with rapturous joy. Since the dog seems eager for a walk, the soldier takes her out, and she leads him happily on her usual evening stroll, right up to the door of a strange house where the soldier's mistress, we gather, is lying in another lover's arms. In each of these stories, man's best friend does him the dubious service of showing him how—and with whom—he's been cuckolded.

KASHTANKA

KASHTANKA, THE EPONYMOUS subject of a short story by Anton Chekhov first published in 1887, is a mutt—"a reddish mongrel, between a dachshund and a 'yard-dog,' very like a fox in the face." She's lived her whole life with a drunken carpenter and his nasty son, both of whom treat her cruelly. The carpenter shouts at her angrily, "Once he had even, with an expression of fury in his face, taken her fox-like ear in his fist, smacked her, and said emphatically: 'Pla-a-ague take you, you pest!'" One day, thanks to the carpenter's carelessness, Kashtanka gets lost in the bleak city streets. Freezing and hungry, she huddles down in the door-way of an inn. "If she had been a human being," the narrator observes, "she would certainly have thought 'No, it is impossible to live like this! I must shoot myself!'"

Fortunately, a kindly stranger adopts her; feeds her bread,

cheese, meat, and chicken bones; and gives her a mattress to sleep on. This stranger, a circus performer, gives her the name Auntie, and introduces her to his other animals: a pig, a white cat, and a goose. When the goose dies, Kashtanka is taught a number of tricks so she can take its place in their performance. A few months later, the sleek and well-fed Kashtanka is in the middle of her circus routine when she spots her former owners in the audience. Gleefully, she leaps from the stage and runs to them, happily returning to her former life of abuse and privation. Before long, "the delicious dinners, the lessons, the circus . . . all that seemed to her now like a long, tangled, oppressive dream."

Like similar stories involving animals, "Kashtanka" is regularly found in anthologies of tales for children. While it's true that youngsters are often fascinated by animals, there's a widespread assumption that "animal stories" are written for the young, as if no mature adult could be interested in a story told from the point of view of a dog. One adaptation of the story available on Amazon, colorfully illustrated, is placed in the age range "8 and up," and the book is summarized as "a quietly sentimental tale of a lost dog." "Children will respond well to the endearing return of a lost pet," notes a reviewer in the *School Library Journal*, and an otherwise enthusiastic reader takes the trouble to warn parents that "one character dies of natural causes midway through."

Yet "Kashtanka" isn't meant for children. It's a painful tale that asks some very difficult questions about happiness and familiarity. Is it instinct that compels Kashtanka to return to her former life? After she first leaves home, trying to get to sleep one night, she recalls with an "unexpected melancholy" the things the carpenter's son used to do to her, including one "particularly agonizing" trick: "Fedyushka would tie a piece of meat to a thread and give it to Kashtanka, and then, when she had swallowed it he would, with a loud laugh, pull it back again from her stomach." Nevertheless, "the more lurid were her memories the more loudly and miserably Kashtanka whined." Why would anyone feel nostalgic for such abuse? Is Kashtanka so deprived that she regards even cruelty as desirable attention? How do dogs think about such matters, if they think about them at all?

Perhaps the familiarity of Kashtanka's former life, despite its privation, provided her with a kind of contentment; I've definitely noticed that once Grisby gets used to something—even if it's something he dislikes, such as taking a bath or getting his nails clipped—he comes to accept it stoically, even compliantly. It's well known that women return again and again to their abusive partners, and prison inmates eventually get used to their daily routines. Children brought up in captivity or isolation often find the outside world threatening, and are drawn to the safety of small enclosures. Perhaps, in some ways, we all prefer what we know.

I suppose it's possible that Kashtanka returns to her original master because he allowed her more independence than the circus performer, who keeps her on her toes with a daily routine of lessons and rehearsals, though we're told that "she learned very eagerly, and was pleased with her own success." Companionship and a daily routine ought to make her happy; in their natural state, after all, dogs live their entire lives within the closely structured social order of their pack, which suggests they understand and even seek out dominance and leadership. In her book *The Companion Species Manifesto*, the cultural theorist and dog agility trainer Donna Haraway, pondering what happiness means for a companion animal, concludes that it is "the capacity for satisfaction that comes from striving, from work, from fulfillment of possibility . . . from bringing out what is within."

If Haraway is right and dogs need to work to their full capacity, then Grisby must be pretty miserable. He doesn't have the opportunity to "strive" every day, perhaps not even to exercise as much as he'd like to, certainly not to satisfy his capacity for "fulfillment of possibility." According to this argument, Grisby, like

Kashtanka, prefers his confinement because it's the only thing he knows. Overweight, neutered, and idle, he's never had the joys of hunting down his own dinner or chasing a female in heat. He's addicted to his captivity, like the Pekingese dogs owned by prostitutes in old Shanghai that acquired opium habits from breathing in the smoke of their mistresses' lamps. Still, there's a kind of arrogance, I think, in the suggestion that a dog like Grisby must be living an unfulfilled life because he's not living to his "maximum capacity." (To be honest, how many of us are?) The argument doesn't take account of his intelligence and flexibility (figuratively speaking—in the flesh, not so much). It seems wrong not to read his displays of excitement and exuberance as indications of pleasure, just as it would be wrong not to read his whimpers and whines as signs of pain.

His repertoire of noises, moreover, is remarkably expressive. His snorts and grunts seem to signal curiosity and contentment, and the other sounds he makes seem equally meaningful. When something's bothering him—when an object's temporarily out of place, for example, or when there's a bag of trash in the room—there's a noise he makes with his lips pursed cautiously that sounds exactly as if he's trying to articulate the word "woof." There's a deep sigh of resignation or satisfaction he makes when settling down for a nap, and a weary inward-outward exhalation that sounds as if he's saying, "*Oy vey!*" When tickled, he makes breathy, openmouthed panting sounds that bear a very close resemblance to laughter, and when he's tired, a series of noisy, cavernous yawns form the overture to his snoring symphony.

In the following situations, I find it difficult to believe Grisby isn't feeling something very close to what we humans call "happiness": when, at the park, I unhook his leash and he goes charging off on the trail of a rabbit or squirrel; when he runs through long, damp grass in the sunshine; when he gets to join human games of Frisbee or soccer; when he chases skateboarders and barks at their boards; when, in spring, he rolls around in fallen blossoms; when he charges up and down with his head in the air; when he noisily wolfs down his dinner; when he bounces down the street on a warm morning, spreading his bulldog bounty; and, not least, when I come through the door on those rare occasions when I've had to be away from him all day. Once, when I came home unexpectedly, he ran around the dinner table three times. Exuberant joy at my return certainly seems the most obvious reading of his behavior, though I must admit, it's also the most flattering. It's hard to escape the common assumption that, emotionally speaking, dogs are easier to understand than people. "Animals are more unrestrained and primitive, less subject to inhibitions of all kinds," Thomas Mann argues, "and therefore in a certain sense more human in the physical expression of their moods than we are." Darwin makes a similar observation. "Man himself," he wrote, "cannot express love and humility by external signs so plainly as does a dog, when with dropping ears, hanging lips, flexuous body, and wagging tail, he meets his beloved master."

Then again, maybe Kashtanka ran around the table three times, too, when the drunken carpenter and his son got home. Maybe that's just what dogs do. Maybe Grisby hasn't had enough experience to fully discriminate between routine and happiness. I wonder: If he were unhappy, would I know it? Would *he* know it? And if we can be unhappy without knowing it, what does it really mean to be unhappy? These are questions it's impossible to answer. However deep my bond with Grisby may be, the absence of a shared language will always come between us. When he clearly wants something and I've tried all the usual things— food, water, treats, petting, a kiss, a scratch, a walk—I look into his eyes, so intelligent and expressive, and it suddenly seems baffling that he can't just tell me what he wants.

French author Maurice Maeterlinck disagrees that we're separated from dogs by their lack of language. His short essay "Our

Friend the Dog," first published in 1903, begins with an epitaph for his "little bull-dog" Pelléas, who has just died at the tender age of six months. From the time he spent with this loving animal, Maeterlinck came to feel that, of all creatures, the dog "succeeds in piercing, in order to draw closer to us, the partitions, ever elsewhere impermeable, that separate the species!" In other words, according to Maeterlinck, the lack of a shared language actually enhances the human-canine relationship. Perhaps it's true. In the absence of language, it's fine for me to kiss Grisby's head, paws, and nose and the insides of his ears; to wash his bottom, clean out his wrinkles, and wipe away his drool. Language would make this kind of thing a little uncomfortable. He could, after all, ask me to stop, which is what children do when they get old enough. Perhaps, rather than translating dog thoughts into human words, we should be wondering, as Alice A. Kuzniar suggests in her book *Melancholia's Dog*, "how to preserve, respect and meditate on the dog's muteness and otherness."

Language, for all its benefits, is also a source of great pain. It's language that brings awareness of time, memory, and death. Language creates the scrim of consciousness that separates us from other beings. Language gives us the capacity to dwell on the past, to anticipate the future. Without language, we'd be living like animals, in the perpetual present, outside time, unaware of death. The impossibility of our returning to this state is something many philosophers have tried to grapple with ("If a lion could speak," according to Ludwig Wittgenstein, "we would not understand him"). In an essay on humanism, the philosopher William James speculates on the connection between language and thought. "To call my present idea of my dog . . . cognitive of the real dog," he wrote, "means that, as the actual tissue of experience is constituted, the idea is capable of leading into a chain of other experiences on my part that go from next to next and terminate at last in vivid sense-perceptions of a jumping, barking, hairy body."

Jumping, barking, hairy bodies have no need to speculate about "the actual tissue of experience"—they're too busy having it. Dogs, unlike philosophers, never doubt. "A dog cannot lie, but neither can he be sincere," claimed Wittgenstein, who clearly never had a dog. Grisby is always sincere, and so was Pelléas, whom Maeterlinck describes in "Our Friend the Dog" as a young bulldog with a generous and gentle nature. Maeterlinck recalls Pelléas, though still a small puppy, "sitting at the foot of my writing-table, his tail carefully folded under his paws, his head a little on one side, the better to question me, at once attentive and tranquil, as a saint should be in the presence of God." To Maeterlinck, a dog's unthinking love and devotion to its master turns human beings into divinities, of whom the dog possesses full knowledge. "I envied the gladness of his certainty," wrote Maeterlinck of Pelléas. "I compared it with the destiny of man, still plunging on every side into darkness, and said to myself that the dog who meets with a good master is the happier of the two."

Clearly, human language is artificial, impractical, and absurd when compared with "a jumping, barking, hairy body," or the solid certainty of a bulldog at your feet. Virginia Woolf wrote of Flush that "not a single one of his myriad sensations ever submitted itself to the deformity of words," yet the author needs "the deformity of words" to describe the dog's life, to translate it into a story we can understand. *Flush* is a book for humans; dogs don't have stories. On this subject, William James, in an essay entitled "A Pluralistic Universe," speculates that "we may be in the Universe as dogs and cats are in our libraries, seeing the books and hearing the conversation, but having no inkling of the meaning of it all." But why assume "the meaning of it all" is something contained in books? Words can't be chewed, chased, or licked; they can't be eaten, growled at, or pissed on; they have no smell; for those of us without language, what use are they at all?

[1 2]

L U M P

LUMP WAS A handsome dachshund owned by David Douglas Duncan, an American photojournalist and a close friend of Pablo Picasso's. The Spanish painter was a well-known animal lover, and he acquired his dogs, according to Duncan, the same way he acquired his women: by "borrowing" them from his friends. If Picasso was a rogue, however, he was also generous—he gave dogs to his dogless acquaintances, partly, he said, to ensure there'd be one around whenever he paid them a visit. Picasso was a man who always needed a dog to hand.

"Lump" (pronounced "Loomp") is German for "rascal" (German names seem to be as common for dachshunds as English names for bulldogs and Chinese names for chow chows—see PERITAS). Duncan acquired the dog from a family in Stuttgart; he was originally intended as a companion for Kubla, Duncan's

Afghan hound, but as it turned out, the two dogs didn't get along. When Lump first laid eyes on Picasso, conversely, it was love at first sight. According to Duncan, in the summer of 1957, when he was visiting Picasso's hillside mansion in Cannes, the one-year-old Lump simply "jumped ship" and spent the next six years living in luxury with his surrogate master. Unsurprisingly—since they were vital to his everyday existence—Picasso owned many different kinds of dogs: terriers, poodles, a German shepherd, a boxer, three Afghans, and a handful of mixed-breed mutts, as well as a goat named Esmeralda. Nevertheless, Lump was the only one he picked up in his arms.

In his book *Picasso and Lump: A Dachshund's Odyssey*, Duncan explains how it was Picasso's habit to work alone in his studio so he wouldn't be distracted; the only one permitted to visit him there was the privileged Lump, who was also allowed to sit at the family dinner table during meals, sometimes standing on the artist's lap to eat from his plate. The dachshund appeared in many of Picasso's paintings, including fifteen of his forty-four interpretations of Velázquez's *Las Meninas*, in which the artist replaced the large dog in the foreground of the original with abstract renderings of his beloved Lump.

In 1957, the year Lump went to live with Picasso, a potential marriage of art and literature occurred when the great-grandson of Victor Hugo, François Hugo, a professional silversmith, arrived in Cannes to deliver two huge platters he had made from designs by Picasso. Hugo's dachshund, Lolita, accompanied the silversmith, and was considered a perfect match for Lump. The two dogs were wed in a touching ceremony, the bride in a veil made from a lace doily, and Picasso promised himself as godfather to the puppies. Despite the dogs' mutual attraction, however, the wedding was unconsummated, and both dogs remained childless.

In 1963, when visiting Picasso at his villa, David Duncan was alarmed to discover that Lump was unwell. The dog had been taken to a vet in Cannes, said Picasso, to be treated for a spinal complaint. Duncan visited Lump at the vet's office and was informed that back problems are to be expected in dachshunds, and that Lump's ailment had no cure. Refusing to accept this unhappy diagnosis, Duncan retrieved the hound and drove six hundred miles to Stuttgart, where he'd acquired Lump seven years earlier. Here he found a dachshund specialist who was willing to try a different kind of treatment, and finally, after several months, Lump made a full recovery (though he was left with a limp that made him walk, according to Duncan, like a drunken sailor). The dog lived for another ten years with his original master, though he continued to visit his surrogate master in Cannes. Their fates seem to have been linked—Lump died on March 29, 1973, and Picasso passed away ten days later.

The curiously shaped creature we call the dachshund is also known as the doxie, wiener dog, sausage dog, and, in Germany, the *Dackel* (although "dachshund" is a German word, it is rarely used in modern Germany). In her essay "Dogs and Their Affections," Ouida describes the breed as "squat, clumsy, deformed . . . as ugly as he is out of place on the cushion of a carriage or a boudoir." Many would disagree; the dachshund is especially favored among artists (perhaps those with a visual mind are more likely to understand the breed's particular charm). Andy Warhol's two miniature dachshunds, Amos and Archie, often accompanied him to parties and gallery openings, and the artist David Hockney, a lifelong admirer of Picasso, paints in the presence of two long and low companions, Stanley and Boodgie.

In the early 1990s, Hockney began to experiment with setting up easels at different heights and putting out canvas, palettes, and paint close to where his dogs usually curled up. When they were in position, he'd rush over and paint them as quickly as possible, and in 1995, he exhibited forty-five paintings of Stanley and

Boodgie in various states of repose. Critics found the images trivial and sentimental, and the exhibition was generally dismissed
as an embarrassment. "Only someone of Hockney's standing and
wealth could indulge in a whim so comprehensively," wrote Jon
Stock in the *Independent*. Nothing Hockney said about the show,
however, suggested he intended it as a "whim." "I make no apologies for the apparent subject matter," he wrote. "These two dear
little creatures are my friends."

The impressionist Pierre Bonnard was another dachshund fan; it's ironic, then, that Picasso denounced the French
painter—using a somewhat canine insult—as "not a painter,
but a piddler." Like his detractor, Bonnard was a *Dackel* aficionado: he owned six sausage dogs in succession, all bearing
the name Poucette (see ATMA). A dachshund is often present
in Bonnard's compositions—usually a small brown creature
with large black eyes. This is hardly surprising, since Poucette
("Little Thumb" or "Thumbelina"), in one incarnation or another, was the painter's constant companion, joining him on his
daily walks, sitting on his knee, and snoozing beside him as he
worked.

Some have suggested the affinity between artists and doxies
may be due to their shared temperament: dachshunds and creative types are both said to be curious, sensitive, stubborn, and
proud. Along with visual artists, writers and musicians are also
drawn to the breed (see QUININE), often said to be the most fe-
line of dogs. Dachshunds lack the tireless energy of larger breeds;
they're not relentlessly social, like many dogs, but are thoughtful
and solitary, content with a single companion, human or otherwise. They apparently quickly grow set in their ways. Some
believe the dachshund to be the perfect writing companion. In
the mid-1880s, the British poet Matthew Arnold returned home
from work every lunchtime to take his doxie Kaiser for a constitutional in Hyde Park, since "he quite expects it, and is the best

of boys." Kaiser died in 1887, and was commemorated in a poem, as was his predecessor Geist, who lived only four years. Predictably, scholars usually dismiss these late elegies as a mawkish and regrettable descent into sentimental "animal verse," best overlooked, perhaps better forgotten.

In "Kaiser Dead," Arnold notes that when he obtained the puppy in London, the dog was "vouch'd by glorious renown / A dachshound true," yet as the creature grew older, it became clear that there'd been a mistake; Kaiser, it turned out, was half collie. This mutt, acknowledges the poet, must have had a different father from his "brother-dog" Max, who, with his "shining yellow coat" and "dewlap throat" was "a dachshound without blot" ("Kaiser should be, but is not"). When Kaiser died, according to Arnold's moving elegy, his half brother Max gazed on his corpse "with downcast, reverent head." Years later, after Max, too, had passed away, Arnold encountered his deceased pet's dachshund doppelgänger in Nuremberg. It was, he wrote, "the darling himself, the same colour, the same sex . . . the same slow and melancholy way. He looked at me wistfully, as if to say: 'I know you, but we must not speak here.'"

While walking Kaiser in Hyde Park, Matthew Arnold may very well have run into the American author Henry James walking *his* dog, Tosca, a small terrier, who, when she died in 1899, was succeeded by a dachshund, also named Max (always popular, the name has become increasingly common since the nineteenth century, and for the last thirty years has been the number one name for male dogs). Henry James acquired Max in spring 1903, when he was living in London; in a letter to a friend he describes his new dog as "hideously expensive . . . and undomesticated," adding that Max boasted "a pedigree as long as a Remington ribbon." In August, he refers to the "poignant drama" of Max's going to bed. By November his new chum is "snoring audibly in the armchair," and by August of the following year, Max, no longer the subject of jokes, is "the best & gentlest & most reasonable & well-mannered as well as most beautiful small animal of his kind."

The English composer and pianist Benjamin Britten owned a dachshund named Clytie, after Clytie Mundy, the singing teacher of Britten's longtime partner, the tenor Peter Pears. In a 1954 photograph of Britten that hangs in London's National Portrait Gallery, the composer is shown holding Clytie to his chest. According to the portrait's photographer, Yousuf Karsh, "the dog demanded to become part of the picture. Britten swiveled on the piano seat to make room for his canine collaborator, who leaped into the safety of his arms, while yet casting a wary eye on me." Doxies are popular with actors, too, and many a Hollywood star had a loyal wiener dog waiting patiently in his or her dressing room (among them: Joan Crawford's Püppchen, William Powell's Schnapps, Alan Ladd's Fritzie, Rita Hayworth's Knockwurst, and James Dean's Strudel).

There's always been a connection between creative artists and dogs of all kinds, not only dachshunds. This tradition is respected at the Maryland Institute College of Art in Baltimore, where Grisby has always been welcome. MICA has a long history of allowing dogs on campus; they're registered at the beginning of the academic year, and wear a bone-shaped identification badge on their collars. Although the policy seems tailor-made for the comfort of homesick students, it's actually a perk enjoyed mainly by the faculty and staff, since however much they might miss their canine pals, few undergraduates have the extra time or money needed to look after a dog while they're away at college. On top of that, animals aren't allowed in the dorms, the cafeterias, the library, or the painting studios; as a result, most of the dogs on campus belong to deans, directors, administrative assistants, and nonstudio faculty members like myself. Anyone who's spent time on campus will have encountered at least one of the college's canine denizens: Cosimo, Budge, Cooper, Ziva, Bingley, Alex, Manchester, Gompers, and, of course, Grisby.

The University of Idaho, MIT, and the University of Illinois at Urbana-Champaign all have pet-friendly dorms; historically, however, keeping dogs in college accommodations has been considered a serious breach of etiquette, leading some students to rebel against the prohibition. When Lord Byron took his bulldog, Smut, with him to Trinity College, Cambridge, in October 1807, he was told that undergraduates weren't allowed to keep dogs in their rooms. Undaunted, he exchanged his dog for a bear. The college authorities asked him what he planned to do with the creature; "he should sit for a Fellowship," Byron replied. The Cambridge dons were less than amused, but since there was no mention of bears in the college rules, there was nothing they could do. Byron was allowed to keep his bear, which, at the end of a chain, accompanied the young aristocrat as he strolled about town.

Not everyone can get away with such audacity. When the young Henry Kissinger arrived at Harvard in September 1946 after serving in the US Army, he smuggled in his cocker spaniel, Smokey, a dog he'd picked up on an army base during the war. After a few months, recalled Kissinger in a talk he gave at Harvard in April 2012, he was called into the dean's office and told that disciplinary proceedings would be brought against him if his dog was seen on campus again. The future diplomat managed to keep Smokey out of sight by boarding him in a local kennel during the day and fetching him back in the evening, after the dean had gone home.

It's a shame more universities don't allow dogs on campus, because in many respects, it's the perfect place for them. Grisby loves coming with me to college; he always enjoys company, and he's never short of friends. Of course, sometimes I have to leave him behind—if he's unwell, for example, or if I have a student who suffers from dog allergies. When this happens, my day changes its entire complexion. From the moment I leave home, everything seems flat and dead, like a meal without salt. My daily routine becomes pointless and dull. I walk directly to work instead of following our usual roundabout path, which—with Grisby's predilections in mind—takes in a selection of interesting smells, a back alley filled with Dumpsters, and a wide area of grass and trees. Without my little sidekick, I'm always slightly preoccupied, and rather than lingering to chat with students or catch up with colleagues, I leave as soon as class is over. Some people might think having a dog in class would be an annoying distraction, but in my case, it keeps me focused and calm.

Still, I need to remind myself that not everyone loves dogs, and the fact that Grisby's thrilled to meet new people doesn't necessarily mean they're equally excited to meet him. I sometimes have Korean students who aren't accustomed to having dogs around, and consider them unhygienic and unpleasant. I also have students who've been raised on farms and have learned to feel detached from animals, to see them as livestock, vermin, or prey. Attitudes toward animals are generally inconsistent. Those who live in countries where dogs are eaten can be totally indifferent to animals bred for the pot, but can still grow fond of a pup that's not destined for slaughter. Pets aren't unknown among farmers, trappers, and slaughterhouse workers, who'll occasionally pick out a special lamb, calf, or runt to bring up by hand. In Western culture, these inconsistencies are perpetuated wholesale. We buy hot dogs at the concession, then shed tears over *Babe* and *Charlotte's Web*; we spend thousands of dollars keeping our elderly pets alive yet happily boil lobsters to death for dinner.

Indifference to suffering comes from living a compartmentalized life. One way to break down these barriers is to welcome pets into places of business, which should happen more often, especially given today's emphasis on environmentalism and sustainability. One day a week, I work in a men's prison, where certain inmates are allowed to take care of a young Labrador retriever for six months (the program is called Canine Partners for Life). Since the first set of puppies arrived at the prison, inmates have apparently become less emotionally isolated and less prone to violence, and have exhibited higher morale. Still, we shouldn't overromanticize animal companionship. The Centers for Disease Control estimates that more than 85,000 Americans are injured each year by tripping over their dogs.

When I was living in California, the graduate school where I worked advertised its campus as "animal friendly," so I was surprised to learn that Grisby wasn't welcome. What "animal friendly" meant, I discovered, was that the campus plantings and structures had been designed to blend in with the environment in such a way as to invite birds, insects, and other animals to make their homes there. Two campus cats, Boots and Herbie, strolled into and out of classrooms and climbed on the tables in the libraries—but my bulldog was verboten.

The campus was about fifteen miles away from home, so I couldn't nip back at lunchtime to take Grisby for a walk, nor could I justify working from home. (Incidentally, it's perfectly acceptable for a woman to say she wants to stay home and raise her children, but try saying you want to stay home to raise your dog.) And while I didn't have to be on campus every day, I sometimes needed to stay well into the evening, and so for the first time since acquiring him four years earlier, I was away from my pal on a regular basis. Of course, I knew he'd be fine—we had a huge yard for him to play in, and David was at home to keep him company. It wasn't Grisby's well-being I was worried about but my own. I'd grown accustomed to having him by my side, at my feet or on my lap; it was wrenching to be without him all day.

The worst part was the terrible little tussle that took place every morning, when Grisby would try to work out new ways to get in the car with me. Whenever he saw me putting on my shoes or heard me jangling my keys, he'd start running around in excitement, assuming we were going out. Sometimes he'd charge through the gate ahead of me and stand expectantly by the car, looking up at me eagerly. I'd have to grab him by the collar and drag him back into the yard. Then I'd lock the gate and tell myself not to look back. If I did, like Lot's wife, I'd be cursed, turned into tears by the sight of his little snout sticking through a crack in the gate, his plaintive whimpers.

For this reason (among others) we returned to Baltimore and to MICA, where Grisby can come to college by my side, where he belongs. This morning—a cold one—we walked to work together, me in my arctic parka and Grisby bundled up against the wind in his blue coat and scarf, eyes shining, ears back. During the break, as usual, students gave him their empty yogurt cartons and cream cheese containers to lick out. After finishing his snacks, he took a short stroll around the room, then put his nose down on his paws and took a nap under my desk, snoring gently. I may not be a painter, but like Picasso and Pierre Bonnard, I need a dog to hand at all times—and in my case, only Grisby will do.

MATHE

THE GREYHOUND IS sometimes called "the dog of kings," and its connection to royalty (see EOS) goes back to ancient times. In his *Illustrated Book of the Dog* (1890), the authority Vero Shaw describes the greyhound as one of the "most delicate breeds of dog" owing to its fragile body, fine skin, and sensitivity to damp and drafts. The mummified remains of these graceful dogs have been found in the tombs of Egyptian pharaohs, preserved beside their masters even in the afterlife. Greyhounds were also fixtures at European courts in the Middle Ages. Dressed in gold collars set with pearls and rubies, they'd accompany royal ladies on their walks; armored in shining breastplates, they'd go out hunting with the lords. In France, greyhounds of a certain breed had the privilege of accompanying their masters whenever they appeared before the emperor Charlemagne. As a sign of their entitlement, these special dogs had their right paws shaved.

The name of the greyhound belonging to England's Richard II is spelled differently in different sources. Sometimes he's called Math, sometimes Matt, and sometimes Match, but most historical authorities give his name as Mathe, which is a diminutive of Mathuin, the Gaelic word for "bear." It was an unusual name for a unique creature: a dog famous for his disloyalty. For many years, Richard apparently doted on this fond and loving hound. Mathe would follow his master everywhere, and when the pair were reunited after an absence, the creature would stand up and press his paws on the king's shoulders. Mathe's affection would have been especially welcome in 1399, when Richard began to lose the support of his people, and faced the threat of deposition from his rival Bolingbroke (Henry of Lancaster).

In August of that year, Bolingbroke's army gained control of the court, and Richard, who'd fled to Ireland, was summoned to meet his usurper at Flint Castle in Wales. During this famous encounter, Mathe apparently left the side of his master, walked over to Bolingbroke, and settled at his feet. The French historian Jean Froissart described the moment in volume four of his *Chronicles*, written in 1400. When the king and Bolingbroke entered the room, writes Froissart, "the greyhound, which was usually accustomed to leap upon the King, left his majesty, and went to the Earl of Derby [Bolingbroke] . . . and behaved towards him with the same familiarity and attachment as he was usually in the habit of shewing towards the King." Surprised, Bolingbroke asked the king what his dog's behavior meant. "The greyhound maketh you cheer this day as King of England," replied Richard, "to which dignity you will be raised; and I shall be deposed."

Betrayal is uncommon in canine lore; Mathe's disloyal behavior is an inversion of the more familiar folklore motif Disguised Man Recognized by Dog, whose best-known manifestation is perhaps the story of Argos and Odysseus. According to Homer, Odysseus's faithful dog Argos, too old to do more than thump his tail, recognizes his disguised master after an absence of twenty years. The dog lives long enough only to share a glance with Odysseus on his return (see ULISSES). The motif also appears in the Celtic legend of Tristan and Iseult. In most versions of this ancient story, the lovers are forced to separate, and Tristan leaves his pointer Hodain with his lady, Iseult. When the hero eventually returns, disguised as a beggar, Hodain recognizes Tristan long before Iseult does. Indeed, so faithful is Hodain that he licks out the cup from which Tristan drinks his fatal poison, preferring death to life without his master.

Richard II recognized Mathe's desertion as an omen of ill fortune; he may have been aware that, at least according to folklore, the behavior of dogs can be a sign of things to come. Black dogs in particular, sometimes with large, glowing eyes, were believed to foretell misfortune. They would appear at night beside burial chambers, gibbets, and other places where death occurs. A black dog was rumored to haunt London's Newgate Prison for more than four hundred years, always appearing before an execution.

When Sherlock Holmes encountered the Hound of the Baskervilles in the late nineteenth century, it wasn't unusual for English locals to believe in a family curse involving a demonic hound. Such beasts have been part of British folk belief for centuries, and each district had its own version of this spectral doom dog. In the north, travelers told of a monstrous black dog with huge teeth and claws known as the Barghest. A similar creature called the Black Shuck haunted the Norfolk, Essex, and Suffolk coastlines. In other parts of the British Isles, this dusky hellhound is known as Hairy Jack, Padfoot, the Churchyard Beast, the Shug Monkey, Skriker, Galleytrot, Capelthwaite, the Hateful Thing, the Swooning Shadow, the Bogey Beast, Old Shook, Moddey Dhoo, or the Mauthe Dhoog. In Yorkshire and Lancashire, the beast was called the Gytrash, and, as Jane Eyre knew, it was said to haunt lonely roads, where it waited to prey on nighttime travelers (see BULL'S-EYE).

An anonymous fifteenth-century German manuscript predicts that "the Devil will come in the form of a black dog," and there are many accounts of demons appearing in precisely this form. Legend tells us that Pope Julius III's legate, the obstinate Cardinal Crescenzio, was literally hounded to death by an avenging spirit in the shape of an enormous black dog, invisible to all but himself. In a verse by French poet and soldier Théodore-Agrippa d'Aubigné, a dark dog appears before the cardinal at the Council of Trent, announcing his impending death and damnation. The same tale is told in "The Cardinal and the Dog," a folk poem by Robert Browning. In Goethe's *Faust*, Mephistopheles first appears to Faust in the form of a large black dog that follows him home. In his depiction of this devil dog, Goethe may have relied on an early chapbook in which Faust is said to own a black dog with demonic powers. (Incidentally, Goethe abhorred dogs; their barking, he said, drove him to distraction.)

Winston Churchill, a poodle lover and the master of Rufus and Rufus II, famously referred to the depression that followed him throughout his life as "the black dog." The phrase was in common currency at the time. Sir Walter Scott, who also suffered from depression, noted in a diary entry for May 12, 1826, "I passed a pleasant day . . . which was a great relief from the black dog, which would have worried me at home." Samuel Johnson, another depressive, observed in a letter to his friend Mrs. Thrale, written in June 1783, that "when I rise my breakfast is solitary, the black dog waits to share it, from breakfast to dinner he continues barking."

Hellish dogs are usually black, but they don't have to be. Scottish locals feared the Cù Sìth, a huge, dark-green dog with shaggy fur and a long braided or curled tail. In Devon, a headless yellow dog called the Yeth Hound or Yell Hound was believed to ramble through the woods at night, wailing. Some of these beasts are distinguished not by the color of their fur but by their distinctive eyes (folklore motifs include Dog with Fire in Eyes and Dog with Eyes like Plates, Tea-cups, etc.). Welsh countrymen share tales of the Gwyllgi ("Dog of Darkness"), a frightful-looking mastiff with baleful breath and blazing red eyes. The Beast of Flanders is also known as Old Red Eyes, and the Canary Islands host the Tibicenas, red-eyed demons in the form of wild dogs covered in long black wool. On mainland Normandy, similar creatures lived in deep caves inside mountains, and were referred to as *rongeurs d'os* (bone gnawers). Dip, an evil black dog in Catalan myth, was an emissary of the devil who sucked people's blood.

Betrayal may be uncommon in canine lore, but it's not unique. In Greek mythology, Actaeon was turned into a stag as punishment for catching a glimpse of the bathing Artemis; he then suffered the indignity of being killed and eaten by his own hunting dogs (the names of all thirty-six hounds are included in Ovid's *Metamorphoses*). To dog lovers, this fate seems particularly harsh; according to folk wisdom, a dog would sooner starve than eat his owner. The truth, however, is clear and stark: dogs are perfectly willing to eat human bodies, and there's no evidence they treat their masters or mistresses differently from any other corpse. Every year, there are numerous incidents of deceased dog owners being partially eaten by their pets. If this fate sounds gruesome, it could be worse; when deprived of food for long enough, dogs have been known to eat their owners alive. Loyalty goes only so far, and karma can truly be a bitch.

Once, I leaned over Grisby when he was dozing, planning to wake him with a kiss. I must have startled him, because to my shock, his head shot up and he went for my neck with a snarl. Luckily, he came to his senses immediately; a moment later, he was wide-awake and ready to play. Still, I'll never forget it.

My best friend took me for a predator and lashed out. I know I shouldn't have taken it personally, but I couldn't help feeling the sting of rejection. That'll teach me to let sleeping dogs lie.

Under the right circumstances, he's probably capable of biting me, but would he feed on my dead body? If I were lying on the floor in a coma, would he eat me alive? It's impossible to imagine, but that doesn't mean it couldn't happen. At some point, even the most devoted love gives way to hunger, and starvation can drive you out of your mind. If it were a matter of survival, I might even find myself eating Grisby—but not without first being transformed by hunger into someone unrecognizable, the human equivalent of the unfamiliar dog that almost went for my throat. I'm reminded of a scene in Charles Maturin's gothic novel *Melmoth the Wanderer*, in which a pair of lovers are locked together in a jail cell and left without food or water. On the fourth night (far too soon, surely), the man leans over his beloved and, in the agony of hunger, sinks his teeth into her shoulder: "That bosom on which he had so often luxuriated, became a meal to him now."

I sometimes wonder whether food isn't really at the heart of my bond with Grisby. These days, since his weight has become an issue, I try not to give him so many treats, though I can't prevent others from doing so. Still, there are some things he loves so much I just have to indulge him now and then. His favorite treats are Frosty Paws, ice cream cups for dogs. Hand Grisby one of these frozen snacks and he'll be preoccupied for half an hour, licking the creamy surface persistently until it starts to melt, at which point his snout gradually disappears into the cup. Frosty Paws are around a dollar apiece, but since a single Frosty Paws keeps Grisby so busy and so happy, to me it seems like a good deal. Of course, they're full of artificial additives—but SweetSpots, the "natural and wholesome" equivalent of Frosty Paws, cost even more.

To those who can barely feed their own children, the idea of ice cream for dogs must seem like a disgusting luxury. Not so very long ago, after all, domestic dogs were fed with table scraps, vegetable rinds, and whatever they could scavenge for themselves. Of course, in those days, dogs weren't kept leashed or penned up and could wander the streets freely, scavenging from Dumpsters and garbage cans or hanging around the back doors of various establishments where food was served.

Grisby gets plenty to eat, but he still forages for scraps on the street, wolfing them down before I can stop him, often before I can see what they are. While it wouldn't be quite true to say he's hungry *all* the time, he's certainly willing to eat at any time (in fact, he's doing so as I write)—but it has to be something more exciting than his usual kibble, which will sit untouched in his bowl as long as he suspects there might be a possibility, however remote, of finding something more exciting. This means he seldom resigns himself to his "official" dinner until both David and I have given every sign of having finished everything we're planning to eat that day.

If people are eating in his presence, Grisby will sit and gaze at them, mesmerized. With me, he'll take liberties, occasionally standing up on his back legs and putting his front paws on my lap, staring at my dinner, sometimes letting out a small whimper. If I'm eating something hot that smells good, he'll lose all self-respect and start drooling; when I'm barefoot or wearing sandals, I'll feel the drops of saliva on my feet. As I get to the end of my meal, Grisby will usually end up with a few scraps, or at least a plate to lick. David doesn't approve of such behavior, and if Grisby comes sniffing around his plate, he's sent packing— though he usually continues his vigil from a distance.

At the end of the day, when there are obviously no more plates to lick, crumbs to catch, or crusts to polish off, Grisby will turn his attention to his kibble. I love watching him eat his dinner—it's a spectacle for the ears as much as the eyes. He snorts, munches, chews, smacks his lips, and munches some more, finishing up with a big drink of water, slurped with noisy gusto. He always gets thirsty in the night, so I leave out a bowl of water at the foot of the bed, and often hear him jump down for a noisy drink, the metal tag on his collar ringing against the edge of the water dish. When he's had enough, he licks his lips, sometimes burps or hiccups, then jumps back up on the bed and immediately falls back asleep with a contented snort.

A fun fact: Grisby's preferred food, above everything else— above pulled pork, Vienna sausage, and barbecue ribs—is white bread. That's right, white bread. He adores it; he can't get enough. Once, at a buffet I made him his own plate of cold cuts, which he never gets at home (David and I are both vegetarian), but, too excited by the smell of the bread, he wouldn't even go near it. His second favorite food is coffee cake, and occasionally, when he's been particularly sweet, I'll buy him his own slice. However, although Grisby surely has his preferences, he'll eat almost anything he's offered. When the pie I was attempting to bake turned into a distasteful, lardy mess, Grisby chomped it up happily, eating my new silicon pie dish into the bargain. As a matter of fact, the joy he takes in his dinner leads me to the conclusion that being eaten by Grisby would be an unusual privilege. For now, however, I'm delighted to be the waiter and not the meal.

NERO

IN DECEMBER 1849, Jane Welsh Carlyle, the wife of Thomas Carlyle, wrote to a friend that she'd recently acquired a little dog. "It sleeps at the foot of my bed without ever stirring or audibly breathing all night long," she wrote, "and never dreams of getting up till I get up myself. It follows me like my shadow and lies in my lap." This affectionate creature, a small black-and-white Maltese mix named Nero, was described by his mistress as "a most affectionate, lively little dog, otherwise of small merit, and little or no training." Her pet made regular appearances in Jane Carlyle's letters and diary entries over the next ten years, and since she was a prolific correspondent and journal keeper, Nero's character and personal habits are chronicled in detail.

We learn, for example, that he was particularly fond of cake, though his everyday dinner was "mostly bread and water (with one spoonful of oxtail soup for relish)." He was washed every day, ran off from time to time, and was stolen by dognapping gangs more than once, but always made his way back home. We also learn that, one morning in 1850, apparently watching the birds, he "sprang from the library window" and, clearing the spiked fence, fell "down on to the pavement." Miraculously, no damage was done. In Virginia Woolf's rendering of the incident (see FLUSH), Elizabeth Barrett Browning's dog, a London neighbor, reports some canine scuttlebutt: "It was common knowledge that Mrs. Carlyle's dog Nero had leapt from a top-storey window with the intention of committing suicide. He had found the strain of life in Cheyne Row intolerable, it was said."

The "strain of life in Cheyne Row" was caused by the fact that both Thomas and Jane Carlyle suffered from fits of depression; Thomas Carlyle, a Scot like his wife, was known to be especially gloomy. Their marriage was childless, and Carlyle's biographer, James Froude, believes it to have been unconsummated. The couple were certainly often at odds, but despite the rumors, their unhappiness doesn't seem to have dampened Nero's vitality. In fact, his antics brought some much-needed variety to Mrs. Carlyle's otherwise dull routine. "My little dog continues to be the chief comforter of my life," she writes to a friend in 1850. "Night and day he never leaves me, and it is something, I can tell you, to have such a bit of live cheerfulness always beside one." Although Jane had many correspondents, Nero was for many years her closest companion, and she was ashamed of the depth of her feelings for him. Admitting that she made "more fuss" over Nero "than beseems a sensible woman," she confessed to her confidante, Mary Russell, "I like him better . . . than I should choose to show publicly."

When Nero first arrived at Cheyne Row, Mrs. Carlyle was afraid that "Mr. C" might find him irritating, and at first this did seem to be the case. Before long, however, she was describing how, when her husband "comes down gloomy in the morning, or comes in weary from his walk," the little creature "dances around him on its hind legs as I ought to do and can't: and he feels flattered and surprised by such unwonted capers to his honor and glory." Before long, Mr. C. was taking Nero for his regular evening constitutional, and though he often referred to the dog as "that vermin," there's obvious affection in his tone. In January 1852, Jane wrote of returning home "to the unspeakable joy of—my dog!"—who, we learn, was so pleased to see his master and mistress that he stood on his hind legs to greet them, a welcome that surely would have brought cheer to the heart of even the sternest Scot.

On November 14, 1855, the couple had Charles Darwin and some of his relatives over to Cheyne Row in the evening for tea. "Dullish all of us," wrote Mrs. Carlyle, "except Nero who dearly likes a tea party—contriving always to coax a *considerable* amount of *cake* out of it." A few weeks later, when the Carlyles were expecting a visit from the critic John Ruskin, Nero was "washed and combed for the occasion," and curled up cozily on a cushion. Jane couldn't help asking Mr. Ruskin what he thought of her dog, and he replied, "I find just one—I can hardly call it *fault*, but one objection to that dog; which is, that one never *can* tell which is head and which is his tail." This kind of affectionate teasing made Ruskin a real favorite with Mrs. Carlyle. In a letter written in 1856, she raved about how the great critic gave her and her husband strawberries and cream on the lawn, "and was indulgent and considerate for even *Nero*!"

In April of the following year, Jane Carlyle was thrilled to discover that Nero's portrait would be appearing in the National Gallery. "Only think!" she wrote to her friend Mary Russell, "my little dog is to have his picture in the Exhibition this year!" The portrait, however, was not of Nero alone; his image had been used in a painting of a nude woman stepping into her bath. At first, Mrs. Carlyle was worried that, since Nero was "recognizable at any distance," anyone looking at the picture would assume the woman was she herself. "The only comfort," she added, "is that naked women in pictures are always *plump*—and I am but two degrees removed from a skeleton at présent."

In 1859, the little mutt was still in good health, swimming in the sea in Scotland with his master (Carlyle, another husband who favored the Cesar Millan approach, admitted to having "flung him into the sea" three times, which apparently gave the dog a taste for bathing). A few months later, however, a tragedy occurred: during a walk to the shops with the Carlyles' maid, Nero was run over by a butcher's cart. Despite a crushed neck and lungs, he managed to hold out through the winter, wheezing and suffering; finally, the doctor eased his pain with a dose of prussic acid. His mistress, naturally, was heartbroken. "My gratitude to you will be as long as my life," she wrote to the doctor who had ministered to Nero so kindly, "for shall I not, as long as I live, remember that poor little dog? Oh, don't think me absurd, you, for caring so much about a dog. Nobody but myself can have any idea what that little creature has been in my life."

By the time of her own death, Mrs. Carlyle had acquired another dog, a pug named Tiny. This dog, too, suffered a traffic accident, which may very well have caused his mistress to relive the terrible scene with Nero six years earlier. According to Carlyle's biographer James Froude, on April 21, 1866, Mrs. Carlyle went for her regular afternoon carriage ride in Hyde Park. Near Victoria Gate she put her pug on the ground, as she usually did, so it could run beside the brougham. "A passing carriage went over its foot," writes Froude, "and, more frightened than hurt, it lay on the road on its back crying. She sprang out, caught the dog in her arms, took it with her into the brougham, and was never more seen alive." After several circuits of the park, the coachman realized something was wrong and, looking into the carriage, saw Mrs. Carlyle "sitting with her hands folded on her lap dead."

No one reading Jane Carlyle's letters and journals could underestimate the value to this sensitive lady of her joyfully devoted pet. Her writing is full of intriguing tidbits about the role of dogs in middle-class Victorian life. Nero accompanied his mistress however she traveled, whether by stagecoach, brougham, or train ("nobody asked for his ticket"). One day he went with her "by two omnibuses to Hampstead," where the pair "spent several hours sitting on the Heath, and riding in a donkey-chair." Apart from his daily dish of bread in oxtail soup, Nero was fed sugar lumps and strawberries, drank tumblers of brandy, and regularly put away more than his allotted share of cake.

Six years before she acquired Nero, Mrs. Carlyle wrote a letter to her uncle complaining about an overindulged dog that sat beside her on a chair at Grange's, an expensive confectioner's in Piccadilly. The gentleman who owned him, she recalled, purchased a "large pound-cake," cut it into pieces, "and handed him one after another" until it was all gone. In the light of London's widespread poverty, she found the scene a little offensive: "It must have been a curious sight for the starved beggars who hang about the doors of such places to see a *dog* make away with as much *cake* in five minutes as would have kept *them* in *bread* for a week." When she obtained a dog of her own, however, Mrs. Carlyle easily outdid the gentleman at Grange's in her extravagance.

England, like most other European countries, has always been more sympathetic than the United States to dog owners. Although I lived in Britain until I was thirty, I never owned a dog during that time, so I didn't realize how dog-friendly my native country could be. According to British law, individual pub and restaurant owners can decide for themselves whether they want dogs on the premises, as long as animals don't have access to the kitchen and food storage areas. When we traveled to the UK,

Grisby was welcome in virtually all the pubs we visited; he even sat beside us while we ate. This was a delightful treat. Normally when we go out to dinner, I have to leave Grisby at home or in the car, which, when I'm so used to having him around, feels disruptive. Personally, I think well-behaved dogs should be allowed in restaurants. It's well known that their presence can help people feel calm and comfortable. Dogs promote civility and conversation; they improve the atmosphere.

In Willie Morris's charming book *My Dog Skip*, which recounts the author's 1940s Mississippi childhood, young Willie's Jack Russell terrier accompanies him just about everywhere—to local shops and restaurants, to movie theaters, to church, even to school. There's no mention of permits, allergies, or health and safety regulations. Reading *My Dog Skip* made me start to wonder when—and why—the United States became so austere about dogs in public places. What I discovered is that laws differ from state to state, but most states have adopted the Food and Drug Administration's prohibition on live animals in retail establishments where food is served, a ban that goes back to 1970.

The question of why this injunction came into effect is more difficult to answer. Some might argue that our immune systems are being challenged less and less in modern life, and as a result, the number of people with dog allergies is rising—but, by the same token, so is the number of people with allergies to perfume, flowers, hair spray, and laundry detergent, none of which are banned from places where food is served. Some might say people are frightened by dogs—but people are also frightened by heights, elevators, open spaces, and flying, and they often learn to accommodate or even overcome these fears. At the heart of the dog ban, I believe, is a falsehood as arrogant and illogical as similar myths perpetuated about human minority groups: that dogs carry disease.

Obviously, there are filthy dogs, just as there are flea-ridden people, but as long as a dog has been dewormed, deloused, and properly groomed, there's no obvious health risk involved in having it around, even if you're eating. If you're in any doubt, I stand as living proof. Over the last eight years, I've kissed Grisby on the mouth hundreds of times. When I've finished eating, he sometimes licks my plate and utensils. I run my fingers through his fur, checking for fleas or ticks; I tickle his belly; we share peanut butter from the jar. When it's time for a bath, he joins me in the tub. If Grisby carried disease, I'd have been pushing up daisies long ago. Yes, dogs do carry bacteria, just as humans do, but dog bacteria are just that—bacteria specific to dogs. When we pick up infections, we pick them up from those with whom we share our microorganisms: namely, other people. As for getting rabies, I suppose it's possible—but you're more likely to be struck by lightning.

Proof of the pudding is the fact that, according to the Americans with Disabilities Act, service dogs *must* be allowed in restaurants, which surely proves they pose no threat to human health, even around food. The act also states that service dogs don't need to wear any particular signage, and those accompanied by such dogs have no obligation to explain what service their dog is trained to perform. These provisions leave the field wide open, one might think, for people with "ordinary" dogs to take advantage of the law, a possibility that often alarms and sometimes infuriates those who rely on their specially trained service dogs in order to live functional lives. To pass off a pet as a service dog, they argue, undermines the laws intended to protect those who are truly disabled. I can't help thinking much of this anxiety relates to the fact that the distinction between a service dog and a "normal" dog is slippery, to say the least.

Managers of chain restaurant franchises, even if they're dog lovers themselves, hardly ever risk breaking the law; after all, if they let an "ordinary" dog enter the premises, outraged customers might complain to the health inspector, who could impose a big fine or even close the place down completely. Alternatively, someone might have an allergy attack and hold the restaurant responsible for the resulting hospital bills. When managers are held to corporate rules, there's no chance of leniency. My local Starbucks won't even let me bring Grisby inside for the two minutes it takes for the staff to hand me a cup of coffee. It wouldn't be illegal—places like bars and coffee shops that serve only prepackaged snacks aren't bound by the same regulations as restaurants—but it would contradict corporate policy. At Starbucks, Grisby's considered a "health violation," though no one can stop me from buying him a hot sausage biscuit and feeding it to him at a table outside. Babies, on the other hand—noisier and more irritating than any dog, and genuine sources of transmittable disease—are welcomed in many restaurants, and are even given their own chairs, tables, play areas, and "changing stations." Grisby never whines, cries, yells, throws his toys across the room, or monopolizes the conversation.

Luckily, there's an easy way out. No one is forced to eat in chain franchises, and there are plenty of independent establishments that, when faced with conflicting and hypocritical regulations, will take their chances and open their doors to dogs. I'm fortunate enough to live within a short walk of three different coffee shops and two bars that actively welcome Grisby (I'd love to mention their names but don't want to cause them any trouble). It's a win-win situation—Grisby gets a few crusts of toast or corners of his favorite coffee cake, and I can eat lunch with a bulldog in my booth. If I sit at the bar, he'll hang out under my stool, licking my legs if they happen to be bare. If I eat at a table, he'll make himself at home underneath, waiting for scraps, clearing up other people's crumbs, then, if we're staying for a while, settle down for a short afternoon snooze. Anywhere that lets dogs join their human friends for coffee and a muffin is a true sanctum of civility. To the proprietors of all such establishments, I salute you.

ORTIPO

IN THE AUTUMN of 1914, seventeen-year-old Princess Tatiana of Russia, volunteering as a Red Cross nurse in a St. Petersburg hospital, grew close to a wounded soldier named Dmitri Malama. After nursing him back to health, Tatiana was delighted to receive a special gift from the dashing soldier: a French bulldog puppy. She named the little dog Ortipo, after Malama's horse. However, on September 30, 1914, writing to her mother, Tatiana realized that other members of the royal family might not be so keen to have a puppy in the palace. "Forgive me about the little dog," she wrote. "To tell the truth, when he asked should I like to have it if he gave it me, I at once said yes. You remember I always wanted to have one, and only afterwards when we came home I thought that suddenly you might not like me having one. But I really was so pleased at the idea that I forgot about everything."

Tatiana's mother was sympathetic; not only did she let her daughter keep the dog, she allowed her to accept a second puppy—another gift from Malama—when the first one died shortly after its arrival. The second Ortipo was a female French bulldog like the first, and she quickly became Tatiana's constant companion. "It is a very cute little thing—I am so happy," she wrote.

Before long, Ortipo was running amok in the palace, raising Cain. Unlike the other Romanov family dogs, Ortipo slept with Tatiana on her bed (her sister Anastasia complained in her journal about being kept awake at night by the bulldog's snores). When she reached her full size, Ortipo received a jeweled collar made especially for her by Carl Fabergé, who also made a series of French bulldog figurines for Tatiana, carved from smoky quartz and decorated with precious gems. The family's letters and journals often refer to Ortipo's adventures; we hear of her chasing Princess Olga's cat through the Winter Palace, getting into a fight with a pig, and giving birth to puppies. ("They are very small and ugly, and who knows what and whom they resemble.") In March 1916, Dmitri Malama paid a visit to the palace; the czarina noted in a letter to her husband, "Ortipo had to be shown to his 'father,' of course."

We last hear of Ortipo in April 1918, in firsthand accounts of the Romanovs' flight, in which Tatiana is described as "struggling to carry Ortipo while dragging a suitcase through ankle deep mud and a howling crowd at the Ekaterinburg train station." Two months later, at midnight on June 16, 1918, Tatiana and her entire family were executed. According to one account, when the revolutionary soldiers went into the house where the czar and his family had been shot, a small bulldog, defending his owners' bodies, barked at them angrily. A soldier stabbed the dog to death, then brought it outside and displayed it impaled on his bayonet, drawing savage cries of victory from the crowd. Finally, the poor creature's body was tossed into the pit along with the bodies of the royal family.

So the story goes, at least. What really happened is a little more difficult to discern. From witness accounts, we know the Romanovs actually took three dogs with them to exile in Ekaterinburg. As well as Ortipo, there were Princess Anastasia's dog, a King Charles spaniel named Jimmy; and Prince Alexei's dog, a springer spaniel named Joy. In the 1990s, the bones of a small terrier-type dog were found alongside the remains of the royal family in the mine shaft in which they'd been buried after their execution, but historians are unsure which of the dogs these bones belonged to, and whether this was the same dog that was allegedly stabbed with a bayonet.

A bulldog impaled on a bayonet is a memorably grisly image; perhaps for this reason, it feels like an embellishment, a vivid detail drawn from folk history. The image tells us what we already believe, or want to believe: our enemies are heartlessly brutal, our noble dogs loyal to the last. In later accounts of military atrocities, however, the story gets more disturbing: the bulldog becomes a baby. During the First World War, British newspapers contained reports of German soldiers tossing babies into the air and letting them land on their bayonets. During the Nanking massacre, Japanese soldiers were said to use enemy babies for bayonet practice, and the same was said of Khmer Rouge fighters in Cambodia. It may well be true that, throughout history, there have been incidents of soldiers impaling babies (and even bulldogs) on their bayonets, but the widespread appearance of this grim motif makes it most likely to be propaganda.

As such, it's related to another theme of historical folklore: the victorious display of a royal dog. In 1648, when Charles I of England was taken captive and held in Carisbrooke Castle on the Isle of Wight, his black-and-white toy spaniel Rogue was imprisoned with him. In January 1649, he was condemned to death for high treason, and on the morning of his execution, the dethroned king, so they say, took Rogue for his final walk across

London's Green Park to the scaffold, which had been erected at Whitehall Palace. According to this version of events, after his master's execution, the king's spaniel was taken up by Oliver Cromwell and toted around London by the Roundheads as the symbol of a fallen monarchy. However, more reliable historians claim that Charles I sent the dog to his wife a few days before his execution.

During their father's imprisonment, Charles I's children were exiled in France, and when his son Charles returned to England in 1660, he brought his favorite toy spaniel with him. It was this son, later to be Charles II, after whom the King Charles spaniel was named. A pack of these small dogs accompanied their master all over the palace, where, like Ortipo, they had free rein, and even slept with the king in his bed. On September 1, 1666, Samuel Pepys attended a council meeting in the palace and wrote in his diary that "all I observed there was the silliness of the King, playing with his dog all the while and not minding the business." Charles II was notably fond of spaniels; still, they had been popular at the British court long before either he or his father took the throne. It was a spaniel, some say, that saved the life of Edward VI in 1549, when his uncle, Thomas Seymour, was alleged to have broken into the king's apartments at Hampton Court Palace in an attempt to kidnap him. His entrance disturbed the dog, which started barking at him. The traitor killed Edward's pet in a panic, but it was too late—the noise had alerted the king's guards. Seymour was later executed at the Tower of London, dying "dangerously, irksomely and horribly," according to eyewitness accounts.

For a very short time, Edward VI was betrothed to the seven-month-old Mary, Queen of Scots. The unpopular engagement was soon broken off, but as dog lovers, the couple might have made a good match. When Mary was executed in 1587, she was said to have been hiding a lapdog in her petticoats, unseen by spectators until after the sentence had been carried out. Some accounts describe this secret companion as a small black-and-white spaniel; others refer to her as a Skye terrier. After the queen's beheading, the blood-splattered lapdog apparently refused to be parted from her mistress's body, and when forcibly removed, she languished away in grief. However, since similar stories were also told of Marie Antoinette and Anne Boleyn, this, too, is most likely an apocryphal tale based on the widespread folklore motif Faithful Lapdog Dies When Mistress Dies.

In fact, Marie Antoinette's spaniel Thisbe (rhymes with Grisby) was also rumored to be a victim of the frightful bayonet. An account of the execution has been widely reproduced: "The Queen's head fell—there was a moment's dead silence— then the loud, agonising howl of a dog. In an instant, a soldier's bayonet pierced its heart. 'So perish all that mourn an aristocrat,' he cried." There are also rumors that, after her death, Marie Antoinette's pug Mops committed suicide by throwing himself into the Seine, and that Anne Boleyn's greyhound Urian was beheaded along with his mistress. (The latter is hard to imagine. Did the dog have its own small scaffold with a dwarf executioner, or were the pair dispatched together, with a single blow?)

Generally speaking, the abandoned dogs of defeated monarchs are better off than those who accompany their master or mistress to prison, exile, or the scaffold. In 1860, during the Second Opium War, Anglo-French forces looting the Chinese emperor's Summer Palace came upon five Pekingese dogs in the otherwise empty suite of a dowager empress. A naval officer rescued two of the dogs, and the Duke of Wellington's brother-in-law also took a pair. The fifth, given the audaciously honest name Looty, was taken by a general, and presented to Queen Victoria. Looty apparently seemed lonely, and so six months later, a second Pekingese was ordered from China, and the breeding of this pair led to a new fashion for Pekes in aristocratic society. Looty's portrait was

exhibited at the Royal Academy in 1862, then hung in Windsor
Castle, where it will no doubt remain until the next revolution.

Fashions in dogs are nothing new. After bullbaiting was outlawed in England (see BULL'S-EYE), the original bulldog was crossbred with the pug to produce a smaller, less fierce bulldog that
could be kept as a pet. Sometimes, the litters of these dogs would
contain a big-eared runt, and the runts were especially prized
by female lace makers, who liked these low-maintenance lapdogs that would keep them warm in the damp English weather,
and draw fleas away from flesh and fibers. Plus, the runts were
small enough for these ladies to keep on their knees or under
their skirts while engaged in their work.

The economic crises of the nineteenth century and the indus-
trial production of textiles caused many of these skilled artisans
to cross the channel, resettling in France, where their skills were
still in demand. Some of these ladies smuggled their toy bulldogs
on board ship with them, hidden inside their clothing. Many lace
workers settled in Calais, but others went on to Paris, where they
lived mainly in the working-class areas of the city, side by side
with the butchers, slaughterers, and meat traders, whose dog of
choice was the native round-faced *terrier boule*, also an accom-
plished ratter. Most people believe it was the interbreeding of
the toy bulldog and the *terrier boule* that gave rise to the French
bulldog, with its compact physique, odd face, and enchanting
personality. This little dog was quickly taken up as a pet by the
Parisian working classes, especially coach drivers, shoemakers,
and other street traders.

The French bulldog has a unique appeal. Aesthetically, other
breeds are undeniably more glamorous and showy, but this bat-
eared beast has a *jolie laide* quality that some find impossible
to resist. Notoriously, these charming little creatures became
fashionable among prostitutes, so much so that during the Belle
Époque in Paris, if a woman was seen walking a French bulldog,

it was read as a sign she was looking for business. Not only did these funny little dogs help attract potential clients; their easy-going character meant they had no problem taking short naps at hotels during the afternoon, when their mistresses were engaged with gentlemen customers. At the time, well-known prostitutes often posed for paintings, postcards, and drawings with their French bulldogs at their feet, in their laps, by their sides, or under their dressing tables, sometimes sporting feathered neckwear or ruffled collars.

Eventually, the dogs became popular with artists, homosexuals, bohemians, and other members of the avant-garde who wanted to show how edgy they could be by having one of these "naughty" animals as a pet. The risqué entertainer Mistinguett owned a French bulldog, as did Madame Palmyre, the woman who owned Toulouse-Lautrec's favorite restaurant, La Souris. Her Frenchie, Bouboule, appears in two of the painter's works: *Le Marchand de Marrons* (1897) and *Madame Palmyre with Her Dog* (1897). Another French bulldog is the subject of *Touc, Seated on a Table* (1879–81). Touc, like Ortipo, had an unusual name; at the time, French bulldogs were usually given names starting with *B*, such as Billie, Bullie, Bébé, Bouton, Boulet, or Bouboule. Two early British champs were Bumps and Bruiser.

The aristocrats of prewar Russia quickly picked up Paris fashions, and the trend for Frenchies was no exception. The Russian craze was started by Prince Felix Yusupov, who, when in Paris as a boy for the Exposition of 1900, saw some French bulldog puppies for sale. He took such a fancy to one of them—a brown pup named Napoléon—that he begged his mother to buy it for him. To his joy, she consented. Felix felt it would be disrespectful to call his dog after such a famous man, so he changed the creature's name to Gugusse. For the next eighteen years, this dog was Yusupov's devoted and inseparable companion, becoming a well-known member of the Russian imperial family. According to the prince (later known as "the man who killed Rasputin"), Gugusse was "a real Parisian guttersnipe who loved to be dressed up, put on an air of importance when he was photographed, adored candy and champagne."

Gugusse slept on a cushion by Yusupov's bed, and appeared in the prince's portrait painted by Valentin Serov (who declared that Gugusse was his "best model"). Purebred French bulldogs have bat ears, like Gugusse in the finished portrait, but in a photograph of Felix and Gugusse posing for Serov, you can see that Gugusse's ears are floppy. Since Yusupov was known for his own vanity, it seems likely he asked Serov to paint his dog with the breed's "proper ears"—which is a shame, as the floppy-eared Gugusse looks interesting and unique.

The famous Russian opera singer Fyodor Chaliapin owned a French bulldog named Bulka, whom he dressed in a coat decorated with little bells on the front; a contemporary said the dog was "unpleasant looking, hoarse, and breathed loudly through her nose." In 1914, the Russian artist Ilya Repin painted Chaliapin reclining with Bulka on a Turkish divan. Disappointed with the result, the artist painted over the canvas in 1917, replacing the figure of Chaliapin with that of his own mistress, Alisa Rivoir, depicted in the nude. One of her arms is wrapped around the charming Bulka, who stayed in the picture.

According to the ship's insurance records, a French bulldog named Gamin de Pycombe went down with the *Titanic* in 1912. The dog's owner, who survived, was a twenty-seven-year-old banker named Robert Williams Daniel; another survivor, R. N. Williams, later recalled seeing a French bulldog trying to keep afloat as the ship went down. Gamin de Pycombe was insured for more than $700, which is quite a sum for pet insurance, even by today's standards. Incidentally, two other passengers on the *Titanic*, Samuel and Nella Goldenberg, were on their way to attend the French Bull Dog Club of America's show at the Waldorf-

Astoria in New York, where Mr. Goldenberg was to be one of the judges. He survived the shipwreck, keeping his appointment at the show.

According to the American Kennel Club, French bulldogs are becoming increasingly common. In 2002, they ranked as the fifty-eighth most popular breed in the United States; in 2007 they ranked as the thirty-fourth, and in 2012 as the fourteenth (the third in New York City), though they're still nowhere near as fashionable as their relative the English bulldog, which places fifth. It's not always a good thing when a breed's popularity increases so fast. Such trends often lead to a rise in puppy mills, backyard breeders, and mixed-blood imports. Plus, some people who buy French bulldogs don't realize that, as a "man-made" breed, they're prone to various medical problems. Because their bodies are similar to those of the English bulldog—whose monstrous head and pushed-in face were achieved by selection for a skeletal malformation known as chondrodystrophy—Frenchies have a number of physical weaknesses, though the smaller dog's problems are, thankfully, less severe than those of its English cousin.

This is not to minimize the French bulldog's potential health problems, which include joint diseases, spinal disorders, heart defects, eye problems, and deafness. Puppies have to be delivered by cesarean section, since their heads are too large to fit down their mother's birth canal. Their short snouts mean that anesthesia is always a risk, and their eyes are subject to dryness and susceptible to injury. Like other flat-faced dogs, moreover, Frenchies don't do well in hot weather. They wheeze, snort, grunt, suffer from labored breathing, experience strange honking fits called "reverse sneezing," and snore like a freight train at night. Their short jaws can cause dental difficulties, and the wrinkles and folds of skin around their eyes, which produce their adorably mournful expressions, can harbor dangerous bacteria.

I fell in love with Grisby at first sight, but I also realize a number of people find French bulldogs unappealing. Lovers of more traditional breeds often find them freakish, and look with distaste on their grossly foreshortened jaws, high foreheads, and protruding eyes. The flat face is supposed to be especially anthropomorphic, designed to appeal to the human "cuteness" response, but many people find it grotesque and disturbing. In fact, when the breed first appeared in Britain in 1898, it caused a genuine scandal. The English bulldog was such a popular symbol of national character that the sight of this miniature French version with bat ears seemed like a mockery of everything the bulldog had come to stand for. The sentiment was summed up in the popular press: "We English, who have always felt a special affinity for our national symbol, must reject this little abomination that has been brought to our country." Fortunately, the feeling was not universal, and the antipathy toward the breed soon dissipated when Edward VII, who ruled from 1901 to 1910, acquired a French bulldog named Peter, whose portrait was painted by famous dog artist Arthur Wardle.

According to the American Kennel Club, "the French Bulldog originated as, and continues to be used as, a companion dog. The breed is small and muscular with heavy bone structure, a smooth coat, a short face and trademark bat ears. Prized for their affectionate natures and placid dispositions, they are generally active and alert, but not unduly boisterous." Under the American Kennel Club standards, the weight of the French bulldog is not to exceed twenty-eight pounds. In general, females range in weight between sixteen and twenty-four pounds, and males between twenty and twenty-eight pounds (my own little abomination is admittedly a few pounds over the limit).

Much as I love the way Grisby looks, I know that he's a product of selective breeding, and his size, shape, and temperament are the result of human choices that might not be in the best interests of a dog. Sometimes I wonder whether he'd be happier if he'd been born with a snout and tail. (I wonder: Does he ever feel sensation in a phantom tail?) Yet it's ridiculous to ask the question "What did nature intend for a French bulldog?" since nature didn't intend a French bulldog at all. Like all domestic dogs, no doubt, he'd quickly perish were he to be released into nature. Grisby was born and bred not for the wild but for the lap. Bully for me that the lap is mine.

PERITAS

ALL WE KNOW for sure about Peritas, the favorite dog of Alexander the Great (356–323 BC), are three short facts relayed by Plutarch in his *Life of Alexander.* We know he was "brought up of a whelp," that he was "loved very dearly," and that, when he died, Alexander "built a city, and called it after his dog's name." This ancient fortification was probably somewhere in India, but as with most of the other twenty cities founded by Alexander— all of which were named after himself, apart from Peritas and Bucephalus, after his horse—the exact site remains unknown.

Plenty of heroic tales have been told about Peritas, but none appear in Plutarch or any other reliable sources. The most persistent of these stories is that the dog saved Alexander's life by biting the lip of an elephant in the Battle of Gaugamela (331 BC), which seems to be a confused version of an incident described by

Pliny that occurred during Alexander's march to India. According to Pliny, the young emperor was given a dog by a king of Albania, but because it ignored the bears and boars presented to it, Alexander had it killed. The same king then sent Alexander a second dog, letting him know that his dogs were unfamiliar with boars and bears and should be tried against lions or elephants. When put to the test, this second dog apparently tore a lion to pieces, then brought down an elephant. However, this champion wasn't Peritas.

We don't even know the breed of the emperor's elusive favorite, whose name derives from the Macedonian word for January, although most historians believe that the dog is likely to have been a Molosser, a now-extinct breed common to the region where Alexander grew up. These dogs were kept and bred by the Molossians, a tribe inhabiting the Epirus mountain region, an area now divided between northwestern Greece and southern Albania. Alexander's tutor Aristotle, in his *Historia Animalium*, praises the Molosser, and notes that this hound, which serves as "guardian of the herds," distinguishes itself from all other dogs by its size and "indomitable courage against wild animals." After Epirus was conquered by Rome in 168 BC, the breed was imported all over the Roman Empire. Virgil says that ancient Greeks and Romans used Molossers as hunting dogs (*canis venaticus*) and livestock guards (*canis pastoralis*). "Never, with them on guard," he wrote, "need you fear for your stalls a midnight thief, or onslaught of wolves, or Iberian brigands at your back."

Many believe Molossers were the ancestors of all the large, muscular, short-muzzled modern breeds: the Newfoundland, the Saint Bernard, and the Rottweiler, along with cattle dogs, mountain dogs, mastiffs, bulldogs, and boxers of all shapes and sizes. Yet the original Molossers were more wolfish and not nearly so muscular as these modern dogs; Peritas would probably have looked more like a large greyhound than a pit bull. In his book

The History of the Mastiff Breed, published in 1886, British dog authority M. B. Wynn writes, "the true Molossian was an erect eared (*altas aure*) slate coloured (*glauci*) or fawn (*fulvus*) swift footed wolfish-looking dog."

Molossers had a fearsome reputation. In Petronius's *Satyricon*, written around the first century AD, the wealthy Trimalchio owns an enormous Molosser, which is brought in on a chain and introduced to his guests as guardian of the house and slaves. This huge beast has one of the first joke dog names in history: Skylax ("puppy"). In *De Spectaculis*, a series of poems commemorating the opening of the Roman Colosseum in AD 80, the poet Martial describes the sight of a deer in the amphitheater being chased by Molossers (since it stopped in front of the emperor's podium, it was given the thumbs up, and left untouched by the hounds). From such fragments, we can get a sense of what Peritas might have looked like (although no doubt many depictions of him were made, none have survived). The best visual representation of a Molosser dates from almost three hundred years after the death of Alexander, in a mosaic found in the vestibule floor of a second-century-BC Roman home in Pompeii, inscribed with the familiar words *Cave Canem* ("Beware of the Dog").

Sometime shortly before 8 AD, the Roman poet Grattius wrote a *cynegetica* (a poem about hunting), of which 541 hexameter lines remain. In these lines, the poet praises the bravery of the *pugnaces britanniae* over the newly imported Molossers—a comparison that suggests that, since the earliest days of civilization, we've always been compelled to rank dogs in hierarchies of our own construction, dismissing some breeds and defending others. Everyone has his or her favorite; most people automatically endorse the breed they grew up with or, like Grattius, the kind most familiar to them (better the *canis pugnax* you know . . .). To a degree, it seems only natural that we should place dogs in categories because we take it for granted that certain breeds have built-in characteristics, of which some are physical (greyhounds are fast, basenjis don't bark, terriers track and chase) and others relate to personality (bulldogs are stubborn, collies smart, spaniels submissive). These characteristics aren't accidental; they come from generations of breeding for selected traits. In Shakespeare's *Macbeth*, the protagonist expresses the same assumption we take for granted today: that there are obvious ways to rank superior, distinctive dogs above ordinary mutts that have nothing special to add to the simple fact of being a dog.

> *. . . hounds and greyhounds, mongrels, spaniels, curs,*
> *Shoughs, water-rugs, and demi-wolves are clept*
> *All by the name of dogs. The valued file*
> *Distinguishes the swift, the slow, the subtle,*
> *The housekeeper, the hunter, every one*
> *According to the gift which bounteous nature*
> *Hath in him closed, whereby he does receive*
> *Particular addition, from the bill*
> *That writes them all alike.*

We should probably bear in mind that Shakespeare was writing in an age when your dog reflected your station in life. The only dogs peasants could afford to keep were curs, mongrels, and strays. Gentlemen kept hounds and water spaniels for hunting, and the aristocracy could afford lapdogs ("shoughs"—see SHOCK) whose only purpose was to be a pet. Such creatures were symbols of affluence and social standing. (Today, in some respects, it's the other way round; adopting a rescue mutt is a sign of social distinction. In these circles, owning a purebred—sorry, Grisby—is considered as immoral and déclassé as owning a blood diamond.)

Our ranking of dogs has serious real-world consequences. Breed-specific legislation calls for the forced neutering of pit bulls, considered "vicious" by nature; if widely implemented, such legislation would lead to the extinction of these breeds (animal rights activists argue that this is a form of genocide). Beagles, owing to their size and passive nature, are used as subjects for laboratory research. *Nureongi* dogs are raised for meat in Korea. Dog shelters are full of mixed-breed mutts waiting to be euthanized, while in 2011 a red Tibetan mastiff changed hands for $1.5 million, making it the most expensive dog ever sold.

The American Kennel Club allows mixed-breed dogs to be shown on the condition that they've been spayed or neutered, a stipulation that would soon put the club out of business if it were applied to purebreds. Of course, in some ways, the very idea of the "purebred dog" is not only morally questionable but also anachronistic (and circular—a dog is said to be "purebred" if its parents were purebred and it "meets the standards of the breed"). For those unfamiliar with showing and breeding circles, the visual demands of a "perfect specimen" seem utterly arcane. Show dogs must be neither beefy nor bossy; they must not have flying ears, a gooseneck, or a pig mouth; they must not be rangy, restricted, rubber-hocked, or rat-tailed. Males must not be bitchy, nor bitches doggy. They must not be weedy or throaty; they must not have pigeon feet nor a pincer bite. They must be neither shelly, skully, nor snipey. The face must be neither flewsy nor cheeky. Certain breeds must have a varminty expression, others a gay tail.

"We may sometimes feel guilty for not just doing rescue, but I think that it is every bit as important to support carefully selected pure breeds," writes a lady named Sylie on a purebred dog chat forum, SpoiledMaltese.com. "We never want to lose the sublime magic of genetic memory and excellence in carefully selected dogs that are true to their breed characteristics, both physical and temperamental." Sylie gives an example: her Maltese MiMi, who "comes from a long line of show dogs," will "sit quietly while I tie her hair up in a pretty top knot." This might sound sweet, but I fear Sylie is a little off track. While it's true

that a sheepdog pup will, for example, exhibit all the character-
istic abilities to herd and control even if it's never seen a sheep
before, these are breed traits, not vestigial memories of a past
life. There's no cause to assume, in other words, that Huskies pine
for snow, that dachshunds prick up their ears when addressed in
German, or that Scottish terriers are most comfortable wearing
plaid (nor, for that matter, that Chihuahuas insist on Taco Bell).
Still, this is what some people apparently believe.

The Pekingese is a good case in point. For the first three de-
cades of the twentieth century, the Peke was by far the most popular pedigree toy dog breed in Britain. To an empire in decline,
these small dogs had a nostalgic appeal, especially since all English Pekes were apparently descended from the five dogs looted
from the Chinese emperor's Summer Palace in colonial Peking
(see ORTIPO). According to rumors, in the Old China of manda-
rins and opium dens, these particular dogs were considered so
valuable they could be owned only by Chinese royalty, and were
never permitted outside the Forbidden City without their own
guard of eunuchs. In London at the turn of the twentieth century, ladies would give Chinese-style tea parties where the guests
wore silk cheongsam-style gowns and sipped delicately from
willow-patterned porcelain. At such gatherings, a well-brushed
Pekingese held on the lap was regarded as a living reminder of
the Imperial Orient. These delicate creatures were markers of
wealth and luxury, and established a long-lived connection between lapdogs and female celebrities.

Clearly, the hierarchies we devise for them mean nothing to
the dogs themselves, who have quite a different set of rankings,
as their mutual encounters reveal. To my eyes, these negotiations
seem almost random, though I'm sure they're governed by a canine logic too complex and subtle for me to understand. When,
during his neighborhood constitutional, Grisby spots another
dog heading in our direction, he'll freeze in his tracks and stare,

seeming to ask himself, "Now let's see, do I submit to him, or does he submit to me?" I can never predict which it's going to be since the answer apparently has very little to do with breed, size, familiarity, or gender.

Although Grisby is a purebred French bulldog, you don't have to go far back in canine history to realize that no breed is really "pure," and French bulldogs are, to put it harshly, a strain of runts. This may be the reason why, despite his certificate from the American Kennel Club naming his mother as Disterhaupt's Polly (David has suggested this may be a misprint for "Folly"), he'll often defer to mutts, lowering first his ears, then his head, then his legs and belly as he sinks to the ground, trying to look as unthreatening as possible. His behavior often surprises me, especially when the other dog is much smaller than he is. When I find myself thinking, "Really, Grisby? That scrawny little mongrel?" I realize I've been making judgments of my own.

Yet in most such encounters, Grisby is the dominant dog, and I must confess there's a part of me that feels—as he seems to feel—that this is only right. Perhaps, to a small degree, I feel as though I'm the one being deferred to, though most of the time I'm careful to separate myself from Grisby. (When people say "I love your dog," I'll say "Me, too," rather than "Thank you"— after all, I had nothing to do with the way he looks.) While it's sweet to see him submit, it's even more enjoyable to watch him show another dog who's boss. My bullyboy will stand bold and stiff-legged, ears alert, the short fur on his back standing up so he looks like a little hyena. His ears go back, his underbite emerges, and sometimes he'll make a low growling sound, like a rumbling stomach. He'll do this to all kinds of dogs regardless of size, from whippets to wolfhounds, though I should add that he's also attempted to dominate a stuffed husky, a bronze statue of a terrier, and a wooden dachshund on wheels.

Along with his unique personality, Grisby has the usual bulldog traits, both good and bad: he's stubborn, headstrong, and willful, but he's also patient and devoted. Still, I'd never realized quite how deep his breed instincts ran until one day, during a walk in the country, we came unexpectedly upon a field of cows. As soon as he saw them, Grisby, who was unleashed, shot off, slipped under the fence, charged right into the middle of the herd, and began to bait them, dancing and jumping, making feints and play bows, stomping, snorting, and weaving in and out of their legs so recklessly that I had to close my eyes, convinced he was going to get kicked. When I dared open them again, he was trotting back toward me, face shining with a proud expression that seemed to say, "See how I gave those creatures a run for their money?"

I've seen him chase cows, horses, and even deer, yet whenever he's surprised by something, he'll jump in alarm. When David received a very large book in the mail, Grisby ran off in fear. Loud noises will also make him bolt, even when they're coming from the television. I'm not sure why—it may be something to do with having eyes at the front of his head rather than at the sides—but he's scared by anything moving on the periphery of his vision. A flag blowing in the wind, a trash bag, a balloon, even the movement of a cushion can be enough to send him fleeing from the room. What a paradox is my bossy little runt!

QUININE

IN 1892, THE Russian author and physician Anton Chekhov was promised two puppies from his publisher, Nikolai Leykin, one of the first breeders of dachshunds in Russia. The author was delighted. But Leykin and the pups were in what was then the capital, St. Petersburg, and Chekhov lived at Melikhovo, about fifty miles south; it was almost a year before they could be delivered. The following April, a servant of Leykin's had a delivery to make in the general vicinity of Melikhovo, and the dachshund pups accompanied him as far as Voskresensk, home of Chekhov's brother Ivan. Here, the dogs were banished to the privy for a week before continuing their journey by cart in the care of Chekhov's sister, Masha, who transported them to Melikhovo, where they arrived half frozen and shivering pathetically.

The pups, a black male and a tan female, soon warmed up.

"The dachshunds have been running through all the rooms, being affectionate, barking at the servants," wrote Chekhov to Leykin, in a letter of thanks the following day. "They were fed and then they began to feel utterly at home. At night they dug the earth and newly-sown seed from the window boxes and distributed the galoshes from the lobby round all the rooms." Masha named the male dog Bromine (Greek for "strong-smelling") and the tan one Quinine (a drug used as a painkiller); Chekhov gave them the nicknames Brom and Khina. In August 1893, after the pups had spent the day chasing hens and geese in the garden, he wrote to Leykin: "Brom is nimble and supple, polite and sensitive, Quinine is awkward, fat, idle, and cunning. Brom likes birds, Quinine digs her nose into the ground. They both love to cry from excess of emotion. I have to smack them almost every day: they grab patients by the trousers, they quarrel when they eat, and so on. They sleep in my room." Interestingly, it was Quinine—the lazy, idle, and potbellied female—who became the author's particular favorite. According to Masha, "every evening Quinine would come up to Anton, put her front paws on his knees and look into his eyes devotedly."

Chekhov's best-known and most anthologized short story features a dog, not a dachshund but a white Pomeranian (also known as a miniature German spitz). In "The Lady with the Little Dog," Anna Sergeyevna's pet gives her lover, Gurov, an excuse to approach her when he first sees her walking by the seafront at Yalta, the resort at which the couple begin their illicit affair. Gurov regards Anna Sergeyevna's lapdog as a sign that she's alone, and may be interested in sexual adventure (though, like most literary lapdogs, it snaps at him when he tries to pet it). Later, desperate to see her again, Gurov visits Anna's hometown but stands on the street outside her house, afraid to knock. When he sees a servant walking the white Pomeranian, he starts to panic. "Gurov was on the point of calling to the dog, but his heart began beating violently, and in his excitement he could not remember the dog's name." For Gurov, the dog is an extension of its mistress; his confusion suggests that this normally cynical man, despite himself, has fallen painfully in love. It's typical of Chekhov's superb restraint that we never do, in fact, learn the name of Anna's dog.

Chekhov's dachshunds Quinine and Bromine produced many litters, and twenty years after they arrived at Melikhovo, one of their grandchildren was acquired by another dachshund-loving Russian author, Vladimir Nabokov. How Nabokov acquired this particular dog is unclear, but we know he inherited his love of the breed from his mother, who, as he recalls in his autobiography, *Speak, Memory*, had a particular fondness for brown dachshunds. "In the family albums illustrating her young years," writes the author, "there was hardly a group that did not include one such animal—usually with some part of its flexible body blurred and always with the strange, paranoiac eyes dachshunds have in snapshots." The dogs in these photographs would have included Loulou, the housekeeper's pet; and Loulou's son Box. In his autobiography, Nabokov recalls Box in his older years lying asleep on the sofa, with his "grizzled muzzle . . . tucked into the curve of his hock, . . . so old and thickly padded with dreams (about chewable slippers and a few last smells) that he does not stir when faint bells jingle outside."

In 1904, when Nabokov was five years old, his father returned from Munich with a dachshund pup that his son, already able to read and write in English, named Trainy, "because of his being as long and as brown as a sleeping car." Trainy—handsome, though apparently bad-tempered—loved to hunt hares and moles in the gardens of Vyra, the family's country estate. When the dog was around eleven years old, in 1915, his hind legs became paralyzed, and according to Nabokov, "until he was chloroformed, he would dismally drag himself over long, glossy stretches of parquet like a *cul de jatte*" (a legless cripple).

After Trainy died, the Nabokov family acquired the grandson of Quinine and Bromide, whom they named Box II (no relation to Basket II—see ATMA). In a line later excised from *Speak, Memory*, Nabokov refers to Box II as "one of my few connections with the main current of Russian literature." This dog, another brown dachshund, followed the Nabokovs into exile in Berlin, and ended his days with Nabokov's widowed mother in a suburb of Prague. As late as 1930, according to Nabokov, "he could be still seen going for reluctant walks with his mistress, waddling far behind in a huff, tremendously old and furious with his long Czech muzzle of wire—an émigré dog in a patched and ill-fitting coat."

Unsurprisingly, dachshunds crop up in a number of Nabokov's novels; searching for them can be an amusing game, rather like trying to spot the cameo appearances made by Hitchcock (another dachshund lover) in his movies. A few examples: in *Laughter in the Dark*, Margot holds a "fat yellow dachshund" in her lap; she "pulled up the animal's soft silk ears so as to make their tips meet over the gentle head (inside they resembled dark pink blotting paper, much used)." In *Lolita*, Humbert Humbert notices a neighbor promenading her "dropsical dackel," and Van Veen, the protagonist of *Ada, or Ardor*, professes an aversion to the breed. At one point in the novel, we are told that a dog entered the room, "turned up a brimming brown eye Vanward, toddled up to the window, looked at the rain like a little person, and returned to his filthy cushion in the next room.

" 'I could never stand that breed,' remarked Van. 'Dackelophobia.' "

This condition is rare among writers, who seem to appreciate canines of all shapes and sizes; indeed, some of the most evocative testimonials to dog love can be found in authors' memoirs. Even when—as was the case with both Chekhov and Nabokov—writing is not their only profession, authors' lives can be lonely, requiring long stretches of introspective, soul-searching solitude.

In such circumstances, the presence of a friendly dog can be a source of quiet encouragement and solidarity (not to mention the perfect excuse to take a break and get some fresh air).

Before the days of laptops, having a dog at one's desk would have been more of a liability than it is today. Small pets in particular can be dangerous when in the proximity of inkwells, lighted candles, and valuable manuscripts. A notorious accident—possibly apocryphal, though all too plausible—occurred around 1693, when the isolated and misanthropic Isaac Newton was working on his calculations in the company of his Pomeranian Diamond, from whom he was apparently inseparable. Allegedly, the pooch knocked over a candle, setting fire to irreplaceable notes on experiments his master had conducted over the course of twenty years. "O Diamond, Diamond," Newton is said to have exclaimed, "thou little knowest the mischief thou hast done!" Yet the pup was forgiven in short order, and shared his master's dinner plate that same evening, as was his custom. Lucky Diamond; it is doubtful the irascible mathematician would have forgiven a housemaid, wife, or mistress so readily.

Diamond is surely not the only dog to have ruined irreplaceable documents. In his memoir, *Bashan and I*, the German author Thomas Mann describes how the wet ink of his manuscripts was often blurred or smeared by the "broad and hairy paws" of his shorthaired pointer, Bashan. Lord Byron must have had the ink of his manuscripts smeared by a number of dogs, and no doubt by other animals, too. Over his short life (he died at age thirty-six), the poet had many unusual pets, including a bear (see LUMP), a fox, monkeys, and various exotic birds such as peacocks and cranes. He also owned plenty of dogs, including a wolf hybrid named Lyon, who enjoyed teasing the bear, and a Newfoundland named Thunder, who was less foolhardy. For a while, the young aristocrat also owned a "beautifully formed, very ferocious bullmastiff" called Nelson, who generally wore a muzzle. When it was taken off, friends recall, "he and his master amused themselves with throwing the room into disorder." Nelson came to an unhappy end. One day in the summer of 1806, he escaped into the stable without his muzzle, sank his teeth into a horse's throat, and refused to let go. Byron's valet, Frank, was forced to shoot him through the head.

Frank would have been particularly familiar with Byron's favorite dog, Boatswain, another Newfoundland, since the pair used to sit together on the box of their master's carriage. In March 1807, when Byron was at college in Cambridge, his friend and neighbor Elizabeth Pigot wrote him a short poetical story called "The Wonderful History of Lord Byron and his Dog," which she illustrated with twelve comic watercolor sketches of her friend and his favorite companion. Byron always called Boatswain the best dog he'd ever had—and he had quite a few. In June 1807, he acquired a part-bulldog puppy, writing to Elizabeth Pigot, "he has already bitten my fingers and disturbed the gravity of old Boatswain, who is gravely di-composed." Three weeks later, he wrote again: "My bull-dog is deceased. 'Flesh of both man and cur is grass.'" On August 11, he wrote asking, "How is . . . the Phoenix of canine quadrupeds, Boatswain? I have lately purchased a thoroughbred bulldog worthy to be the coadjutor of the aforesaid celestials—his name is *Smut*!"

In November 1808, "old Boatswain" died at the young age of five, after contracting rabies. Byron, unafraid of being bitten and becoming infected himself, wiped away Boatswain's slaver with his own hands, nursing the dog until the disease took its toll. Upon his companion's death, the grieving lord composed the lines "Epitaph to a Dog" (also sometimes referred to as "Inscription on the Monument to a Newfoundland Dog"), which he had carved on the elaborate memorial he built for Boatswain's tomb at Newstead Abbey, the family estate. "Near this spot are deposited the Remains of one who possessed Beauty without Vanity,

Strength without Insolence, Courage without Ferocity, and all the Virtues of Man without his Vices."

His loss was tragic, yet a month later, Byron was writing to a friend asking for greyhounds—"as many more of the same breed (male or female) as you can collect." While living in Italy, he acquired a bulldog, Moretto, and two mastiffs; in Greece he adopted another Newfoundland, named Lyon, after his wolf hybrid. This second Lyon had the sad duty of accompanying his master's body home to England after Byron's death at Missolonghi. Byron, in his will, requested no ceremony or funeral service, no inscription on his tomb apart from his name and age, and that Boatswain's body should not be removed from its vault. By this time, Byron had sold Newstead Abbey, so his own body was buried in a nearby churchyard. Boatswain remains at Newstead; if you ever visit, you can pay your respects at his grave.

Dogs played a central role in the life of the novelist Sir Walter Scott, as well as in his work. Scott; his wife, Charlotte; and their five children owned many breeds over the years. The author's two favorites were reportedly Maida, a cross between a wolf and a deerhound; and Camp, a large and handsome bull terrier. He also had two greyhounds, Douglas and Percy. An article published in 1877 in *Chambers's Journal* describes how "Scott kept one window of his study open, whatever might be the state of the weather, that Douglas and Percy may leap in and out as the fancy moved them."

The dogs owned by Charles Dickens included two mastiffs named Turk and Linda; Sultan, a cross between a Saint Bernard and a bloodhound; a terrier named Mrs. Bouncer; a Newfoundland named Bumble; and a shaggy white creature called Snittle Timberry, described by some as a Havana spaniel and by others as a terrier. Under his original name, Timber Doodle, this small dog was a gift to the author from the comedian William Mitchell during Dickens's first visit to America in 1842. Timber Doodle was brought back to London and renamed Snittle Timberry, after an incidental character in *Nicholas Nickleby*—this new name was felt to be more sonorous and expressive, though the pooch was usually called Timber for short. Though Timber failed to mate, being considered "weak in the loins," he lived to be very old, and accompanied Dickens and his family on all their trips, including a visit to Italy, where he experienced a terrible case of fleas. The only cure was to shave off all his hair. "It is very awful," wrote his master, "to see him slide into a room. He knows the change upon him, and is always turning round and round to look for himself. I think he'll die of grief."

Virginia Woolf and her husband, Leonard, owned a cocker spaniel named Pinka, a gift from the writer Vita Sackville-West. Leonard Woolf, the founder of the Hogarth Press, often allowed Pinka to join him in the office, and one of his employees recalls the sight of Mr. Woolf, usually very businesslike, "making an entry in a ledger while Pinka calmly climbed on a chair and licked his nose." On another occasion Pinka apparently devoured a set of Leonard's proofs. Heartbreakingly, the morning after the young spaniel's sudden death, Virginia came upon Pinka's inky paw prints on her blotter; they caused her a sudden, horrible "stab of grief."

Some authors find it difficult to maintain a balance between inner and outer life. For these reclusive types, dogs can provide a useful bridge between life on the page and life in the real world. Emily Dickinson's retracted nature was tempered by a brown Newfoundland, a gift from her father; the dog accompanied her on the long walks she enjoyed in the woods and fields of Amherst. The poet referred to this creature, Carlo, as her "mute confederate," and he was her closest and dearest friend. Emily Brontë, another loner, was followed on her walks across the moors by a dog named Keeper, who began as an aggressive guard and developed into a quiet companion; when his mistress died, Keeper was observably grief-stricken.

Dogs like Carlo and Keeper are life-changing. Before I had Grisby, I'd heard people claim their dogs had changed their lives, but I always assumed it was an exaggeration, a figure of speech. When I was planning to acquire a dog of my own. I wasn't sure I wanted my life to be changed, but it has been, and in ways I could never have anticipated. Since I've had Grisby, it's easy for me to see how you could sacrifice anything to spend time with your dog, or to make your dog happy. Luckily, most dogs are easy to please, and all Grisby seems to want is for the two of us to be together.

He loves going out with me, but if I stay home, I get the impression he'd rather be indoors with me than outdoors with anyone else. He doesn't mind the fact that we're not "doing anything," as long as we're together. I've written most of this book curled up on the couch with Grisby snuggled between my calves and thighs, giving the occasional snort of content. When I write sitting up, he'll lie by my side with his head on my hip; if he falls asleep in that position, he'll slide onto his stomach and snooze with his nose on his paws. At other times, he'll sit at my feet and chew on his latest bone or tug toy (according to the tug toy hierarchy, he is officially an "aggressive tugger"). From time to time, I'll rub my bare feet against the warm, velvet folds of his ears and belly, and he'll respond with a companionable nuzzle or lick. If he's not right at my feet, he'll be lying close by, his paws tucked under his chin, resting on the ground in a pose of placid resignation.

When it comes to dog aesthetics, I'm naturally drawn to the Grisbyesque: sturdy, squat, and muscular, rather than floppy and elongated. Yet thinking about dachshunds has made me realize that I know at least four women—all middle-aged, all divorced or widowed, all either writers or academics—who are utterly devoted to their miniature dachshunds. In order to help me understand the phenomenon, I e-mailed Marion, to ask what she loves so much about her sausage-shaped chum. Her response was immediate and effusive. "Dachshunds are incredible to sleep with as they burrow under the covers and function as a hot water bottle," she wrote. "They are great to cuddle because they are just the size of a human baby. They can fit right in your lap when driving and hook a paw out the window. They have a big dog personality as far as fetching and pride and fearlessness go . . . not at all a toy or purse pooch." Funnily enough, these are the very same things I say about Grisby. Perhaps dog breeds are not really so different as we like to think—or perhaps, because we love our own dogs so much, we can't help believing they're unique.

ROBBER

RICHARD WAGNER WAS one of the great dog lovers of musical history, yet despite his professed devotion, the German composer could be a thoughtless and forgetful master, and he had a history of losing his dogs. He misplaced his first companion, a brown poodle, somewhere in the city of Magdeburg. Then, in the early years of his first marriage, he adopted a young wolf cub to appease his wife, Minna, who was pining for a child. However, as Wagner puts it in his autobiography, *My Life*, "this experiment did not increase the comfort of our home life," so a few weeks after being adopted, the pup was "given up."

Wagner met his most famous dog, Robber, at a shop in Riga owned by an English merchant. For some unknown reason, this handsome Newfoundland became so attached to the musician that whenever Wagner came to the merchant's shop, Robber would follow him home. Even after Wagner and Minna had moved, the dog would loiter pathetically outside their former residence. The animal, according to the composer, "so touched the hearts of the landlord and my neighbors by his fidelity, that they sent the dog after me by the conductor of the coach to Mitau, where I greeted him with genuine effusion, and swore that, in spite of all difficulties, I would never part with him again."

This all sounds very noble, but if Robber had known what he was in for, he might have chosen to stay with the English merchant. In 1839, the composer and Minna, who was pregnant, had to go on the lam to avoid the demands of Wagner's creditors. Thanks to friends who bribed the Cossack guards, Wagner, Minna, and Robber managed to escape across the border to Prussia, where they hid in a smuggler's shack while Wagner tried to arrange for them to stow away on board a small merchant vessel sailing to London. The journey to the coast was a nightmare. Robber was too big to fit in the carriage, so the shaggy dog had to run beside the vehicle all day long in the blazing sun. At last, Wagner, "moved to pity by his exhaustion, and unable to bear the sight any longer," hit upon "a most ingenious plan for bringing the great animal with us into the carriage." Whatever this plan was, its ingenuity may be judged by the fact that the coach quickly overturned, causing Minna to miscarry, and the threesome had to take refuge with a peasant family until she recovered.

Wagner needed to travel by sea because he had to evade the authorities; however, in *My Life*, he claims he did so out of consideration for Robber. It's difficult to see why the dog would have been more comfortable at sea than on land; as it turned out, this journey, too, was a nightmare. Robber had to be hoisted laboriously on board ship, then concealed belowdecks with his human companions for what was supposed to be an eight-day voyage. Owing to bad weather, the crossing took not eight days but three and a half weeks, during which time the threesome all suffered from terrible seasickness. To make things worse, Robber took an "irreconcilable dislike" to the sailor who was bringing them food, and "flew at him with renewed rage each time he came climbing down the narrow steps." This precarious journey inspired Wagner's opera *The Flying Dutchman*, which premiered in Dresden on January 2, 1843.

Finally, after weeks of sickness, uncertainty, and discomfort, Wagner and his companions arrived in London, where they crammed themselves into a cab, Robber lying "crosswise from window to window," and drove to a boardinghouse in Soho. The composer, who spoke no English, began negotiating for a room, and somehow, during the general confusion, Robber disappeared. Wagner speculates that he "must have run away at the door instead of following us into the house." The distraught couple spent the next two hours searching for the creature they had spent so much trouble smuggling abroad. Finally, after great anxiety, Robber strolled nonchalantly into the house. Wagner learned afterward that the dog had been seen almost a mile away in Oxford Street. "I have always considered his amazing return to a house which he had not even entered," remarks the composer, "as a strong proof of the absolute certainty of the animal's instincts in the matter of memory."

The couple's final destination was Paris, which they reached in September 1839 by carriage ("my efforts to hoist Robber on top being attended by the usual difficulties," notes his master), and it was here that the dog was lost for good. Wagner claims his companion was stolen, but he also admits that constant anxieties over money had frayed his temper, and he'd been taking out his frustrations on Robber (his dog whip is displayed at the Wagner Museum in Lucerne, Switzerland). It would be nice to think the dog left of his own accord; let's not forget it was Robber who "adopted" Wagner in the first place. Perhaps he found a French master to bond with, and simply followed him home.

Wagner had one last sighting of his lost companion—an event he describes in some detail. About a year after the dog's disappearance in Paris, his poverty-stricken master set out to return a borrowed metronome, and the first thing he saw in the dense morning fog was Robber, whom he took initially for a ghost. But the dog was real. Wagner called out to him; Robber approached with caution. Wagner moved toward him with outstretched hands, but this sudden movement "seemed only to revive memories of the few chastisements I had foolishly inflicted on him during the latter part of our association, and this memory prevailed over all others." Robber took off, and Wagner, still clutching the metronome, chased him through a labyrinth of indistinguishable streets until he lost sight of his dog altogether, "never to see him again."

Unlike most dogs, at least in literature (see KASHTANKA), Robber seemed far from overjoyed to encounter his former master again. I can barely imagine how painful it would be to lose Grisby, then find him again, only to have him run away from me in fear. But in Wagner's telling, the episode is not painful, and what prevents it from being so is the musician's reaction. He's not disconsolate and regretful, as one might expect, but angry; any sympathy he feels is not for the dog but for himself. "The fact that he had fled from his old master with the terror of a wild beast filled my heart with a strange bitterness," he proclaims. Later, in Dresden, he wrote a long story about Robber, "An End in Paris," in which a dastardly Englishman (Wagner despised the English) steals a handsome Newfoundland from its master, a man who loves the dog so much he once sold his own waistcoat to buy it food. The tale, it seems, is an attempt to appease the author's guilt and regret at his treatment of this well-traveled dog.

A series of smaller dogs followed Robber, many of them appropriated from his friends (like Picasso, Wagner had a lifelong tendency to "borrow" other people's dogs). There were Peps,

Minna's spaniel; another spaniel named Fips; Pohl, his landlord's hound in Geneva; Kos, his second wife's fox terrier; Branke, a Saint Bernard; and Marke, another Newfoundland, who outlived his master. Among all these dogs, the only one to succeed Robber in Wagner's affections was Russ, also a Newfoundland, acquired after the musician had finally achieved wealth, social status, and a villa in Bayreuth. Like Robber, Russ was supremely protective of Wagner, only this time the dog was kindly treated and remained with his master until the end. In fact, on the day Russ died, Wagner was scheduled to travel to Vienna to give a concert; he delayed the trip for a day to ensure his companion was properly interred. Like Byron's Boatswain, Russ was buried in the tomb that had been prepared for his master, in the garden of Wahnfried, Wagner's villa. The epitaph on the headstone reads: "Here Lies and Watches Wagner's Russ." When he died eight year later, his master was buried in the same tomb.

Richard Wagner's second wife, Cosima, recorded more than four hundred of her husband's dreams, many of which involve the dogs that were central to his life. In one, Wagner is riding in a carriage with Russ, whose leg is injured, only instead of sitting on the seat beside his master, Russ is squeezed underneath the bench, a place into which he would never fit in real life. In the dream, Wagner has lost his hat, and looks back from the carriage to see the hat pursuing him, rolling behind, as if to catch up with him, "like a little dog." There's also a dream in which a greyhound comes and places its paws on Wagner's chest, and a dream in which Marke has the power of human speech and sings an Italian duet with a stonemason's apprentice. The composer also dreamed about clubbing Peps to death and hiding his dead body under a pile of bricks. In a more auspicious dream, Russ was fatally run over by a carriage but restored to life by his master's kiss.

Many musicians have shared their lives with special dogs, though few of these relationships have been as complex and conflicted as Wagner's. For example, among the pieces written by Rossini during his retirement in Paris is a work entitled "Love Song to my Dog." Rumor has it that Chopin's Valse op. 64, no. 1, known both as the "Minute Waltz" and as the "Little Dog Waltz," was written after the composer saw his terrier puppy chasing its own tail. In the 1890s, the musicians who gathered around the Italian composer Ferruccio Busoni were known as the Leskovites, after Busoni's omnipresent Newfoundland, Lesko—not to be confused with Levko, Rachmaninoff's spaniel, whose presence, he declared, "was more agreeable company than the society of human beings." The quality of discriminating sagacity is especially valued in the Newfoundland, described in Vero Shaw's *Illustrated Book of the Dog* of 1890 as "docile, and always sagacious and faithful to his master . . . his fine discriminating intelligence soon distinguishes the friendly visitor, and bids him welcome."

The cellist and conductor Mstislav Rostropovich was sometimes mocked for his devotion to his dachshund Pooks, described in the press as "the effete miniature dog who accompanies Rostropovich everywhere he goes." When he was exiled from Russia in 1974, the musician flew to London first class, accompanied by two cellos, one suitcase, and Pooks, who had to stay six months in quarantine; Rostropovich interrupted his English tour to visit his small chum in the infirmary (see YOFI). He readily confessed that Pooks was his prize possession and favorite companion, declaring, "I love Pooks more than anything else in the world." When his master moved to Washington to direct the National Symphony Orchestra, Pooks apparently developed a taste for cocktail parties and other social events, where he was always the center of attention.

The composer Erik Satie wrote several works "for a dog," and once said to his friend Jean Cocteau, "I want to write a play for dogs. I have the set design already—the curtain rises on a bone."

Satie would no doubt have approved of Laurie Anderson's *Music for Dogs*, a twenty-minute high-frequency composition performed outside the Sydney Opera House in June 2010. The canine audience certainly seemed to appreciate the piece, much of which was imperceptible to human ears, wagging their tails and barking in encouragement.

My favorite musical work inspired by a dog is the eleventh section of Elgar's *Enigma Variations*, known officially as Variation XI (*Allegro di molto*), "G.R.S.," and, more colloquially, the "bulldog variation." This is a stirring piece, and the story behind its composition is equally inspiring. The "G.R.S." in the title refers to Elgar's close friend George Robertson Sinclair, the energetic organist of Hereford Cathedral, though the piece has nothing to do with organs or cathedrals, and only little to do with G.R.S. "The first few bars," explained Elgar, "were suggested by his great Bulldog, Dan (a well-known character)." One day, when Elgar and Sinclair were taking a walk beside the River Wye, the organist threw a stick for his dog, and in pursuit of it, Dan fell down the steep bank of the river. He was.apparently a strong swimmer and paddled upstream to find a landing place. "Put that to music," suggested Sinclair. So Elgar did. In the first bar of the piece, violins portray the bulldog's tumble; his movement through the water is suggested in bars two and three by basses and bassoons. In the second part of bar five, the full orchestra conveys his arrival on dry land; the horns, winds, and lower strings describe his triumphant bark. A carved wooden statue of Dan now sits on the bank of the River Wye by King George Fields, commemorating his adventure.

Elgar was so fond of Dan that, in George Sinclair's visitors' book, he would often jot down notes and musical illustrations inspired by what he referred to as the "Moods of Dan," ideas that later became motifs in his work. These included "Dan uneasy," "Dan triumphant," "Dan wistful," and "the Sinful Youth of

Dan." A mood noted in April 1898—"He muses (on the muzzling order) *piangendo*"—was later used as the "Prayer" motif in *The Dream of Gerontius*, whose manuscript (initials: DoG) has the word "DAN" written at the top. The composer liked to include these kinds of personal codes and references in his work. When Dan died in 1903, Elgar made a last entry among the "Moods of Dan"—a scale of E-flat major, or "Es" in German ("S" for Sinclair). The scale is missing one note, the D (for Dan) under which Elgar has written the single word "alas!"

Sinclair and his dog were famously inseparable. Dan not only sat by his master's side in the organ loft but also attended all rehearsals when Sinclair was appointed conductor of the Birmingham Festival Choral Society in 1900. One of the society's members drew a caricature of the conductor and his bulldog— the sketch is entitled *The Metamorphosis of Dan*—which depicts Sinclair on his podium and Dan on the floor by his side; but the heads of the two have been switched, so the orchestra is being led by a bulldog in a tuxedo, with a human-headed canine at his feet. This cartoon reminds us that it's the head, not the torso, that determines who—or what—we "are." As artists like William Wegman have long been aware, a dog dressed in human clothes is still, endearingly, a dog; in the same way, any creature with a human head, however bizarre, is regarded as human. Beyond this, the sketch makes us uncomfortably aware that the dog's expression of affection (sitting at someone's feet all day, gazing at him or her in mute adoration), if manifested by a human being, would be considered not love but delusion and pathology: the behavior of an obsessive stalker.

I practice the piano every day, and when I do, Grisby sits underneath, at my feet. Occasionally, he actually sits on the pedals, making it difficult for me to use them, but I'm happy to forgo those functions in exchange for the special pleasure of resting my feet against a bulldog's bottom. I like to imagine Dan sitting

beside Sinclair in the organ loft, or sitting at the base of the conductor's podium; it makes me think about Grisby's relationship
to space. He, too, has the charming habit of making himself at
home in such places, the borderlands and thresholds of interior
space—under chairs and tables, under my office desk, in the well
of the front car seat, and similar dog-friendly nooks. When I wear
a long dress, he likes to sit under my skirt—not hiding exactly,
just camping. Early in our relationship, when Grisby was very
small, I sometimes made the awful mistake of kicking him or
stepping on him, oblivious to his silent presence at my feet until
I stood up and pushed back my chair. I soon learned to look down
before I stand up, as I can be sure that if Grisby's not on my lap
or by my side, he'll be sitting somewhere at my feet.

He does have a doghouse of sorts—a collapsible canvas kennel, like a small tent—that sits on the floor by our bed, but for
many years he rarely went into it of his own accord. In his mind,
apparently, it was the place he was sent to for punishment when
he was in trouble, or when we went out for the evening and left
him behind. These days, however, he seems to have made peace
with his tent, and I've discovered that if I roll up the front flap so
the doorway is open, he'll often retreat there of his own free will.
This happens regularly now. He'll fall asleep on our bed; then in
the night he'll jump down for a drink of water and won't jump
back up again. Although he's still strong and healthy, leaping
back up onto the bed when he's half asleep is obviously sometimes not worth the trouble, and sometimes in the morning I'll
find him snoring happily in his tent.

Collapsible canvas tents are one of the many new alternatives to the old-fashioned wooden kennel that was once felt to
be perfectly adequate for the outdoor dog. These days, instead of
(or as well as) a wooden kennel, your dog can have his own log
cabin, igloo, castle, glass temple, or Tudor mansion. He can even
have a luxury villa with front porch, staircase, balcony, sun-

room, four-poster bed, wall-to-wall carpeting, and webcam—
and a climate that can be controlled by your iPad (freeze him
out if you see him fouling the rug!). Such amenities, it strikes
me, are toys designed to amuse the dog's owners rather than
the dog. Still, the fact that they exist at all suggests we have a
strong need to imagine our dogs are "at home" in our absence;
in other words, to ensure the process of projection continues,
magically, when we're not around. If, while you're away, you
can think of your dog hanging out on his porch or taking an afternoon nap in his four-poster bed, it helps ease the separation,
reassuring you that he's still your happy dog and hasn't reverted
to his animal nature just because there's no one around to see
him as Max, Toby, or Jake.

According to the author of *How to Raise and Train a French
Bulldog*, before dogs became domesticated, the pack leader
would always sleep in a high place, and by letting your bulldog
sleep on your bed, you're giving him the idea that he is your
equal. "It is best that your bulldog's place of relaxation be a
basket on the floor," he continues, "as this will prevent possible
behavioral problems associated with dominance." If that's how
dogs judge sleeping arrangements, then I should be the one
sleeping on the floor. Grisby's not my equal; he's superior to me
in almost every way—certainly in terms of temperament, pa-
tience, fortitude, sociability, openness, and self-acceptance. Occasionally, however, rather than sitting *at* my feet, Grisby will
sit *on* my feet, just as he likes to sit *on* the piano pedals. He's
in an archetypally submissive position, yet it's a subtly dominant maneuver, both exerting ownership and, at the same time,
making it difficult for me to get up. I found nothing about this
paradoxical pose in *How to Raise and Train a French Bulldog*,
but I did find it mentioned in *Bashan and I*, Thomas Mann's
memoir of life with his dog. When Bashan takes up this station,
Mann finds it "quaint, cosy and amusing to feel him sitting

upon my foot and penetrating it with the feverish glow of his body." The posture's symbolic meaning, to Mann, is irrelevant; what matters is the feeling it stirs up in him—a feeling I know very well. "A sense of gaiety and sympathy fills my bosom," writes Mann, "as always when I am abandoned to him and to his idea of things."

S H O C K

IN HIS SPEECH listing the ranks of different dogs (see PERITAS), Macbeth mentions "shoughs." These were curly-haired lapdogs; the term was a Tudor coinage, and is now obsolete. Some believe the word was originally Icelandic, while others speculate that it came from an Old Norse variant of "shag" (as in shaggy-haired); either way, the Nordic etymology suggests the dogs were probably miniatures of the spitz breed, much like the modern Pomeranian. These popular pets were also known as "shock dogs" (owing to their shock of fur, often trimmed into a fashionable "lion cut"), and just as sheepdogs today are sometimes called Shep, shock dogs were sometimes named Shock.

This is the name of a lapdog belonging to Belinda, the comely and cosseted heroine of Alexander Pope's mock-epic poem "The Rape of the Lock," which, in keeping with its satirical style, presents Shock not as an individual in his own right but as a collection of various clichés about lapdogs. We first meet Belinda's pet in the poem's first canto, when the heroine's morning ritual is described. She is still dozing at noon, when "*Shock, who thought she slept too long, / Leapt up, and wak'd his Mistress with his Tongue.*" The sexual suggestion here would come as no surprise to eighteenth-century readers; it was common for love poets to regard lapdogs as little rivals, nestling gleefully on their mistress's lap or between her breasts or thighs, the fortunate recipients of sexual favors permitted to no human suitor. In Henry Carey's "The Rival Lap-Dog: A Song," the poet-lover complains that his mistress never grants him a single caress, yet Dony, her lapdog, "*May Kiss without measure, / And surfeit himself with the Bliss.*" Poet Jonathan Smedley, in "On the Death of a Lap-Dog," recalls jealously how his mistress's pet, another Dony, would lie upon her "*downy breast*" and idle smugly on "*her sleep-inticing Lap.*" Isaac Thompson, in his poem "The Lap-Dog," enviously watches his beloved fawning over her pet: "*Securely on her Lap it lies, / Or freely gazes on her Eyes; / To touch her Breast, may share the Bliss, / And unreprov'd, may snatch a Kiss.*"

In visual art, this stereotype of women goes back to the fifteenth and sixteenth centuries, when small dogs represented both female inferiority and bourgeois decadence, which often took on erotic overtones. In the Renaissance, it became fashionable for portrait artists to paint wealthy ladies in the company of their lapdogs. Over time, the inclusion of these pets came to suggest a certain kind of sexual freedom; eventually, the depiction of a complacent creature in a lady's lap came to seem naughty, even lascivious, especially when the lady in question was a famous courtesan. (Examples include Lorenzo Costa's *Lady with a Lap Dog*, Girolamo Forabosco's *Lady with a Dog*, Bronzino's *Portrait of a Lady in Red*, and Francesco Montemezzano's *Portrait of a Woman*.) Some considered it distasteful that paintings of respectable matrons should include what had come to be seen as a symbol of coquettish idleness; in 1582, Cardinal Paleotti wrote a decree condemning the inclusion of lapdogs in women's portraits—he regarded the creatures as indecorous, silly frivolities that undermined the sitters' dignity. The cardinal was no doubt especially incensed by Titian's *Venus of Urbino*, in which the naked beauty's silky little dog is painted in exactly the same fleshy tones as Venus herself. According to art expert Julian Mitchell, the dog in Titian's painting represents the lady's "politely concealed pubic hair," a sleight of hand suggesting that the intimacy between the subject and her pet borders on bestiality. Is this the unspoken taboo that underlies our anxieties about ladies who are inseparable from their dogs?

Lapdogs that fulfill sexual needs make men seem superfluous, which is clearly another source of anxiety. "The Rape of the Lock" is a satire, but like all effective satire, it touches a nerve. Through Belinda, Pope mocks the wealthy beauties of Restoration society who aspire to do nothing more than "Dance all night, and dress all Day," and who lack any kind of moral grounding. In canto 2, Ariel, the spirit watching over Belinda, senses a "dire Disaster" in the air. He fears she may be about to lose her virginity or crack a china jar, "stain her Honor, or her new brocade." Will she "lose her Heart, or Necklace, at a Ball," or has Heaven decreed "that *Shock* must fall"? Since this last seems to be the worst of all possible horrors, Ariel sends his lesser elementals to guard Belinda's body and accessories, assigning himself to be the "Guard of Shock." This irony recurs in canto 3, when the theft of Belinda's hair leads to an uproar comparable to the horror that's felt "When Husbands or when Lap-dogs breathe their last."

The mock-elegy was a Restoration tradition; many such poems were written on the death of a mistress's lapdog. In John Gay's "An Elegy on a Lap-Dog," the poet laments the death of his lady's dog, another Shock. "No more thy hand shall smooth his glossy

hair," writes Gay, "And comb the wavings of his pendent ear."
As is common in the genre, the poet concludes his verse by ad-
juring his beloved to find a more suitable companion. "In man,"
he advises, "you'll find a more substantial bliss, / More grateful
toying, and a sweeter kiss." In his jeu d'esprit "On the Premature
Death of Cloe Snappum, a Lady's Favourite Lap-Dog," Dr. Anthony Fothergill of Bath imagines the body of the deceased Cloe
being put to use: "Now Clo's soft skin—dear, precious stuff! /
Adorns fair Delia's fav'rite muff." Again, the image of a woman
caressing a furry animal in her lap has obvious sexual connota-
tions; the fact that the creature is deceased raises the additional
question of whether the lapdog is an individual or an ornament,
a pet or a piece of property.

If men seem superfluous in the presence of lapdogs, children
often do as well. When these toy pets came to be associated with
mistresses and courtesans, they also came to suggest unfulfilled
maternal instincts. In John Caius's volume *Of Englishe Dogges*
from 1570, the author refers to little dogs that "satisfie the delicateness of dainty dames and wanton womens wils," and points
out, "these puppies the smaller they be the more pleasure they
provoke . . . to succour with sleep in bed, and nourish with meat
at board, to lay in their laps and lick their lips as they ride in
their wagons." He concludes, "This abuse peradventure reigneth
where there hath been long lack of issue, or where barrenness is
the best blossom of beauty."

Most literary lapdogs, like Shock, are the living accessories of
indolent women. Ladies, in the male imagination, love to wash,
dress, and feed their fawning fur-babies, bedecking them from
snout to tail in ribbons and ruffs. These spoiled creatures are
given mittens to protect their paws, porcelain teacups to drink
from, and satin sheets for their tiny beds. By implication, the ladies who lavish attention on them are not to be trusted. They are
weak, fickle, and frivolous—a stereotype that was already well

grounded by the Middle Ages, when loving your lapdog to excess was a sin that could lengthen your time in purgatory. In *The Canterbury Tales* Chaucer's flirty, foolish Prioress brings her lapdogs with her on a holy pilgrimage, feeding them cake and milk. In Shakespeare's *King Lear* the crazed monarch imagines he's putting his selfish daughter Goneril on trial, and breaks down in despair when even her lapdogs turn against him.

While their sexual use is mostly a male fantasy, lapdogs did, in fact, serve practical as well as emotional purposes. They attracted fleas away from their mistress, kept her hands and feet warm in winter, alerted her to danger, and kept her home free from vermin. Different breeds have been fashionable at different times and places. Terriers were popular in the Middle Ages; spaniels were the lapdogs du jour for ladies of the Tudor and Stuart aristocracies. In England, the "Spaniel Gentle" was the most popular ladies' lapdog until the Restoration, when Charles II fell for the breed. Suddenly, it rose in dignity and status, changing from a lapdog to a gentleman's companion, and acquired a more masculine name: the Cavalier King Charles spaniel. Shock dogs remained in vogue until the seventeenth century, when William III ascended to the British throne in 1689. The new king arrived from the Netherlands accompanied by a number of pugs dressed in orange ribbons—the breed had recently been brought to Holland from China—and flat-faced pets were all at once the must-have lapdogs for society ladies. After Joséphine married Napoléon, her pug Fortuné wore a leather collar with two little silver bells on it along with a name tag on which were engraved the words: "I belong to Madame Bonaparte." Joséphine apparently refused to kick Fortuné out of the royal bed, despite her husband's requests, so the three of them slept together. Napoléon could hardly complain—before their marriage, Fortuné had carried concealed messages to and from his mistress inside his collar while she was confined at Les Carmes prison, since only the pug had visiting rights. Was it Fortuné's own Napoleon complex, I wonder, that led him to meet his Waterloo in a scuffle with the chef's English bulldog?

In her novel *Mansfield Park*, Jane Austen uses Lady Bertram's pug to comment not only on this wealthy matriarch's laziness and lack of backbone but also on her failure as a mother. Lady Bertram, whose four children have grown up, treats Pug like a baby. He's always in his mistress's lap or in her arms, and when the family goes for a walk in the garden, Lady Bertram soon grows exhausted from "sitting and calling to Pug, and trying to keep him from the flower beds." Apart from these occasional languid strolls, Pug and his aristocratic mistress laze on the sofa all day long, idle and decadent, rising only for tea. To drive the point home, Austen explicitly condemns Lady Bertram for "thinking more of her pug than her children." Similarly, in Virginia Woolf's *Mrs. Dalloway*, when having tea with her mother, Elizabeth Dalloway sits her dog, Grizzle, on her lap and feeds him on scones. Observing the scene, her mother suddenly realizes how immature Elizabeth is; clearly, "she cared for her dog most of all."

Alexander Pope, the author of "The Rape of the Lock," was himself a dog lover. His own pet was no lapdog, however, but a female Great Dane named Bounce, immortalized in Jonathan Richardson's painting *Alexander Pope and His Dog, Bounce* and Pope's poem "Bounce to Fop," a satirical address purportedly from Bounce to the king's mistress's lapdog, ridiculing the manners of dogs at court. Bounce and her master were inseparable. The dog had the run of Pope's villa in Twickenham and took her meals from a golden dish. His relationship with Bounce turned Pope into an animal advocate; he wrote letters to the newspapers to complain against the cruel treatment of beasts, and strongly opposed the dissection of dogs in scientific experiments. When Bounce had puppies, Pope sent one of them to the king's eldest son, Frederick, Prince of Wales; the dog's collar was engraved with the famous lines: "I am His Highness' Dog at Kew; / Pray tell me Sir, whose Dog are you?"

Pope, it's said, never left his house without the company of Bounce—and with good reason. Because of a form of tuberculosis contracted in his youth, the poet had a stunted, hunchbacked body. As an adult, he was only four feet six inches tall, and he was always frail. He had a lifelong fear of venturing into public places, not only because people gawked at his unsightly form, but also because his widely published lampoons had made him many enemies and he was easy to recognize. Bounce served as protection (and, some might say, compensation). Pope's small size would have made this huge dog seem even larger. For the same reason, in Karel van Mander III's painting of Raro, the Great Dane of King Frederick III of Denmark, the dog is shown standing beside the Italian court dwarf Giacomo Favorchi. (Perhaps this is also the inspiration behind the popular Animal Planet show *Pit Boss*, in which Shorty Rossi and his team of little people rescue and rehabilitate aggressive dogs.)

If I were attacked, would Grisby protect me? Although I've never had the chance to find out, I never feel unsafe as long as he's with me, even in the grittier areas of Baltimore. The young men hanging out on the corners are always curious about him, coming over to ask me what kind of dog he is, how much I paid for him, and whether he's "up for stud." Yet perhaps it's foolish of me to feel safe. When David, walking Grisby in the dark, fell and badly injured his shoulder, Grisby came over and looked at him in a quizzical way, he later said, but otherwise seemed unconcerned. I had the same experience when I slipped and fell in the snow. Grisby came over and sniffed the ground beside me, perhaps recalling the half-eaten bagel he once found not far from the spot. This does not seem auspicious. According to *How to Raise and Train a French Bulldog*, the breed displays "average intelligence," but in Grisby's case this sounds like grade inflation.

Rather than protecting me from danger, he once even led me into it. I was walking him off leash in the park one morning when he picked up an interesting scent, turned a corner, and disappeared into the bushes. Assuming he was on the trail of a rabbit or fox, I called after him; suddenly, a man leaped out of the foliage performing a series of mad kicks and chops. He was angry and unkempt; when he started to yell obscenities, I realized he was either drunk or mentally ill. His tirade was incoherent, but the gist of it was clear: Leash your dog, lady!

In that part of the park, people always let their dogs off leash; still, the freedom is unofficial. Legally, the man was in the right, and I had no grounds to complain. Still, his obscenities seemed uncalled for, as did the five-inch knife he took out and flashed at me as I went to clip on Grisby's leash. Later, I wondered whether I should have gone to the police. Had Grisby simply disturbed a homeless man sleeping off last night's binge, or had he ferreted out a maniac lurking in the bushes? Oddly enough, the following week, I read about a similar incident in J. R. Ackerley's *My Dog Tulip* (see TULIP). Ackerley describes how, when he is walking with Tulip through the London neighborhood of Brook Green, they pass a homeless man, who cries, "Don't let that dog near me! They ain't to be trusted!"

"You don't look especially trustworthy yourself," responds Ackerley, and adds, "I might be thought to have hit a nail on the head, for he at once fumbled with a jack-knife out of his miscellaneous garments and, opening it with some difficulty, flourished it after me."

According to the author of *How to Raise and Train a French Bulldog*, the breed can be trained in protection and attack, but its main proficiency lies in "bite work," which means goading, baiting, and stubbornly hanging on. I've definitely seen this skill in Grisby. Sometimes, when we play, he'll cling so persistently to a rubber ball that I can lift it up off the ground and he'll hang there in the air, swinging by his teeth like a cartoon bulldog. He can be territorial, too, and is disinclined to share. He's been banned from Charm City Dogs day care for "guarding the water bowl" and "trying to get into it" with smaller dogs. We can't buy treats or toys for our cats because Grisby will grab them; he also likes to take on the cats themselves, chasing one and doing his best to hump the other, but they've both learned to shrug him off; they seem to find him slightly ridiculous. Even with other dogs, he's hardly much of a threat. He'll growl and grumble, and his fur will stand on end, but I've yet to see him actually cause any trouble. He is, as the British say, all mouth and no trousers.

TULIP

TULIP WAS A female German shepherd owned by J. R. Ackerley, a gentle man of letters who spent most of his quiet life living in London, where he worked for the BBC. Ackerley was well into his fifties when he acquired Tulip, and in this ebullient animal, the distant Englishman encountered the ideal friend he'd been searching for all his life. Their story is told in Ackerley's best-known book, *My Dog Tulip*.

Tulip originally belonged to the author's working-class boy-friend Freddie, who was sent to jail during the course of their relationship. When this happened, Freddie's hostile, uneducated family took in the dog, and Ackerley, distressed by the heart-breakingly miserable conditions in which she was being kept, decided to rescue her—a story told in his novel *We Think the World of You*, in which the dog goes by the name of Evie (see DOUCHKA).

In real life, she was called Queenie; Ackerley decided not to use her real name in his books after it was suggested to him that "Queenie" might cause a certain amount of titillation—the author had been an outspoken member of London's gay community for some time (in 1956, this was not the kind of attention one wanted to attract, especially if one worked for the BBC). In *My Dog Tulip*, Ackerley presents himself as a natural loner, and Tulip, despite her erratic, aggressive, and often infuriating behavior, provides him with the satisfying and loyal companionship he'd never managed to find with another human being. The fifteen years he spent with Tulip were, he said, the happiest of his life.

Some might describe Ackerley's relationship with Tulip as codependent, even pathological. It is certainly unusual. For example, Ackerley wants his dog to enjoy all the pleasures existence has to offer, including sex and motherhood. As a result, Tulip is not "altered," and her master seems especially fascinated by the changes her body undergoes when she's in heat. His energetic attempts to find her a suitable mate constitute the most lurid part of the book. The author takes Tulip along to meet a variety of potential partners: handsome, heavy Max; a nervous charmer named Chum; the aristocratic Mountjoy; and Watney, a "rather wooden terrier, with a mean little face streaked black-and-white like a badger." As Tulip comes into heat, Ackerley begins to grow rhapsodic about "this exquisite creature in the midst of her desire," and describes eager suitors sniffing at her gleaming vaginal folds. Much to his dismay, however, Tulip rejects them all in favor of "a disreputable, dirty mongrel" named Dusty, who gives her the clap.

When Tulip gives birth—here again, no detail is spared—Ackerley plans to drown her puppies in the bath, but finds himself unable to do so. Quickly, Tulip grows bored with them, and her master needs to get them out of his flat; as a result, they're handed over to various acquaintances and strangers. Contemplating the potential suffering of the little creatures he's given away in such a cavalier manner, Ackerley meditates philosophically about the intertwined lives of men and dogs. "I realized clearly, perhaps for the first time," he writes, "what strained and anxious lives dogs must lead, so emotionally involved in the world of men, whose affections they strive endlessly to secure, whose authority they are expected unquestioningly to obey, and whose mind they can never do more than imperfectly reach and comprehend."

My Dog Tulip is the rare dog book in which the dog truly takes center stage. Indeed, so central is Tulip to the narrative that there's very little about her we don't know (and some readers will no doubt feel they know far too much). There is a great deal of information, for example, on Tulip's toilet habits. Ackerley explains how much he loves to watch his dog "at stool," and how those buried in the graveyard Tulip uses for her morning dump should be honored to have such a fine young beast leave offerings on their graves. ("Are not its ghosts gladdened that so beautiful a young creature as Tulip should come here for her needs, whatever they may be?") More disturbing to the modern sensibility is Ackerley's regular failure to clean up after his pal—except when commanded to do so by an angry greengrocer whose shop doorway Tulip spectacularly befouls. In another scene, as the dog squats to take a dump on the sidewalk, an irate cyclist yells at her owner, "What's the bleeding street for?"

"For turds like you," Ackerley shouts back.

Like all subjects, as *My Dog Tulip* reveals, attitudes toward dog feces have a history. In England, dog laws have been around since 1791, but those regarding defecation are more recent, going back to the 1970s in most places. Still, I don't remember anyone scooping when I was a kid. There was more dog poop around, naturally, and yes, you occasionally stepped in it. (It's supposed to be good luck!) "Dog mess" was simply another hazard of city life, like construction, cigarette smoke, and exhaust fumes. If you got some on your shoes, it was nasty, but the world didn't come to an end. There were no Dogipots©, no Doody Danglers©, no Poop-Freeze© aerosol spray. Today, signs everywhere urgently remind us: Dog Waste Is a Threat to the Health of Our Children! Dog Waste Degrades Our Town! Dog Waste Transmits Disease! Leash and Clean Up After Your Dog—It's Required by Law! (In England, the signs are usually more concise: "No Fouling.")

I once read a letter in the *New York Times* from a self-righteous dog owner who wrote that, when walking his dog, he carried a bottle of water with him so that, as well as scooping, he could flush away his dog's urine rather than allowing it to pollute the city streets. I rolled my eyes at the idea; others, it turned out, felt similarly indignant. A follow-up letter the next week suggested that this dog owner, although certainly well meaning, was misguided; sniffing other dogs' urine is a major part of every city dog's life. Most dog owners have some sense of how their pet's world is limned and circumscribed by the smell of their own urine and that of their fellow canines. To a dog, urine is a special code, a means of conveying information and communication; it can identify friends and enemies by sniffing their traces on the street. Washing away your dog's urine, according to the letter writer, is rather like closing down your dog's Facebook page. We might dislike the fact that Man's Best Friend uses the streets as a toilet, but while to you it's urine, to your dog it's a map, a logbook, a journal, and the source of endless curiosity and pleasure. Leave it be.

In *My Dog Tulip*, Ackerley makes one thing very clear about dogs: bathe and dress them all you want, but they still shit in the street. For many people, especially those who've never had a dog, having to deal with animal feces on a daily basis is unthinkable, no matter how sweet the creature it comes from. One of the more surprising things I've learned since acquiring Grisby is that dog poop is a very touchy subject. Quite literally, some can handle it, and others can't. Some find even the thought of it highly objectionable (as if theirs doesn't smell). There are even people who seem unable to bear the sight of a dog doing its business. They turn their eyes and gaze elsewhere, anxious for it to be over as quickly as possible. They look horrified if the subject comes up in conversation, wash their hands obsessively after any canine contact, and turn pale if they glance a turd in the street. By the same token, some owners seem to have difficulty pronouncing the words "shih tzu." The website allshihtzu.com gives us the correct pronunciation, making it clear that "the word 'shit' is not in this word," adding defensively, "Note: We do wish to make it clear that if 'shit' *were* in this dog's name, we could certainly say so."

"Don't get me started on the canine species," ranted a colleague recently by e-mail. "Every time I look out my window, some anorectic Yoga mom is kneeling on my front lawn, gloved hand poised to catch Fido's heroic deposit, still warm from the orifice." What, I wonder, is at the root of this anxiety? The same people who find dog feces so disturbing are perfectly happy to discuss their own "regularity" at great length, even at the dinner table. Parents of new babies, in my experience, seem to talk about little else. When it comes to the toilet training of human beings, there's an entire industry of books, charts, step stools, and booster seats designed to encourage the process, including motivational stickers boasting slogans like "No Accidents Today!" and "Yay! I Pooped!" As Colette Audry points out in *Behind the Bathtub*, as soon as you acquire a vested interest in another living thing, "you have to start taking cognizance of such matters: proper functioning of the bowels suddenly assumes great importance." Yet discussion of the subject seems socially acceptable only if the living thing is a human being.

Bulldogs, of course, are notoriously stubborn. This natural inflexibility, on top of Grisby's "average" IQ, meant that we spent the first year of his life taking him outside five or six times a day.

To make matters worse, we live on the fifth floor, so every trip involved waiting for the elevator. If it was an emergency, and if the elevator took a while to arrive, the result was not pretty. I still have memories of standing shivering on the corner of the street in a wind so cold it made my eyes stream, waiting impatiently for Grisby to "answer the call of nature."

Happily, as soon as he got the hang of it, Grisby began to take great pleasure in his daily dump, and as soon as he did, I, like Ackerley, began to enjoy watching him. I realize the subject may be distasteful to some, but I love everything about my dog, and that includes his toilet habits. I especially enjoy the fact that he's so refreshingly open and up front about it. Sometimes—in his own way—he's even dignified. If we're not in a hurry, he takes his time finding the perfect spot, the right pile of wet leaves or street corner, the perfect paving stone or patch of lawn. Like most dogs, he has what are known as "substrate preferences." He likes long, damp grass; wet soil, sand, or undergrowth—anything that lets him kick up his heels and spread his stink around. Once he's found the right spot, he turns his sturdy little body around and around, sniffing to make sure he's made the right choice. Finally, he starts to shuffle his back legs into position like a gymnast gearing up for a tumble; then his big ears go back, his snout goes up, he strains and tenses his back legs like a March hare, and out comes his neat brown pile.

Most of the time, I use the newspaper method, slipping a section of the previous day's *New York Times* under his rear as soon as he picks a spot, though I also carry plastic bags in my purse in case he's caught short. I try to be conscientious about it, and I usually do a good job; still, I'm surprised how presumptuous strangers can be. People have no problem tapping me on the shoulder and pointing out that I've "missed a bit." A friend was recently in the immigration line at JFK when a blind man's guide dog had an attack of diarrhea; airport officials had no hesitation asking the dog's owner to get down on his hands and knees to clean it up. I've personally witnessed people yelling angrily from passing cars, shouting and honking their horns at those miscreants who fail to pick up after their dogs.

No sympathy is spared for the poor dog owner. Just yesterday, for example, Grisby took an unexpected dump about twenty feet from the doorway of our building as we were returning from the hardware store. I put down my two cans of paint, moved my purse and keys to the other arm, dutifully folded the newspaper over Grisby's deposit, then took it in one hand while I picked up the cans of paint in the other, along with my purse, keys, and Grisby's leash. As I reached the steps in front of our building, a neighbor, outside to smoke a cigarette, waved hello. Instinctively, I waved back—using the hand clutching the loosely folded newspaper. A mistake. As I returned the man's greeting, a small but conspicuous piece of feces flew out and, as if in slow motion, described a perfect arc, landing right at his feet. I saw his eyes follow it to the ground; I saw him look up at me expectantly, but at the same time, Grisby was pulling me up the steps and into the revolving door. I did not resist this forward momentum. A more civic-minded person would, no doubt, have returned outside after putting down her burdens to retrieve the errant turd, but I let it lie.

I feel no guilt. Picking up after Grisby is always a complicated balancing act, and I do the best I can while juggling leash, keys, purse, and newspaper—and sometimes with book bag, laptop, coffee, and even groceries. Still, people are fast to complain; either I've done an inadequate job, or they don't like the spot he's chosen (it never seems to cross their mind that every spot on a city sidewalk is outside some building or other). Some people even have signs in front of their homes showing a picture of a squatting dog with an *X* superimposed, or a "witty" slogan like "If your dog needs to poop, please not on our stoop" or "No pee, no poop, no problem." "Do you think that tree was put there just for your dog to use as his personal toilet?" asked a passerby in the street last week, as Grisby was settling down for a dump. "Can't you take him somewhere else?" But where *is* a city dog supposed to go? "Curb Your Dog," say the signs, but in my neighborhood, either the streets are lined with parked cars, or they're dangerously busy; it's hard enough for a pedestrian to be seen, let alone a small squatting bulldog.

The man's complaint was revealing. I had my newspaper ready to catch the deposit, so he knew I wasn't going to leave it under the tree. What he was objecting to, it seems, was simply the sight of a dog taking a dump on the street, where strangers are forced to watch. (Perhaps he thought dogs shouldn't be allowed to defecate at all.) It's true that fecal residue can be unsightly, and that people might slip on it or get it on their shoes; still, not long ago, the streets were full of horse manure, and people survived. Plus, nothing compares to the pollution caused by that most toxic of all animals, human beings. My neighbor may have been offended by a misplaced piece of turd, but this pales in comparison with the contamination caused by his cigarette smoke. Dog feces, especially from a small dog like Grisby, are nothing compared with human-generated toxic waste, air pollutants, oil spills, or chemical leaks, not to mention ozone, lead, traffic fumes, and everything else that poisons the atmosphere.

The issue was discussed at length in a recent article on the subject published in the journal *Critical Public Health*, the significance of which may be undermined by the authors' decision to use the term "poo" (the British version of "poop"), perhaps the most vapid of the dog owners' many euphemisms. But what *is* the acceptable usage? This was another conundrum I had to deal with on acquiring my first dog. The most frequently used terms all sound wrong to me—either too infantile ("poop," "go potty"), too prudish ("go to the bathroom," "answer the call of nature"), too clinical ("evacuate," "empty the bowels"), or too evasive ("take a walk," "step outside"). "Take a dump" is the term David and I tacitly agreed upon, which, if not exactly genteel, is at least accurate and expressive.

The study in *Critical Public Health* addresses the "symbolic representation of 'dog poo' " in a London neighborhood. Through a series of interviews with local residents, the authors conclude that the "sense of disgust evoked by 'dog poo' " is disproportionate in relation to the "actual incidence of dog foul," a paradox that suggests people's complaints about dog poop have far more to do with aesthetics and superstition than with public health. As the authors of this article explain, anxiety about "dog foul" is symbolic. The chance of picking up an infection from dog feces is lower than ever, yet people still react to feces with horror. A perfect representation of social rot, dog feces are an imaginary embodiment of other problems, less easily defined or articulated: incivility, estrangement, public vandalism, and environmental neglect. In other words, we're the ones who are fouling our own neighborhoods—then laying the blame on our dogs.

ULISSES

ULISSES WAS AN eccentric mongrel acquired by the Brazilian author Clarice Lispector after her divorce, when her sons had left home. "I needed to love a living creature that would keep me company," she told an interviewer. "Ulisses is mixed-race, which guarantees him a longer life and greater intelligence. He's a very special dog." Elsewhere she confessed, "I and my dog Ulisses are mutts."

In her children's story *Almost True*, Lispector's smart mongrel introduces himself in his own voice. "I'm a little impolite," he admits. "I don't always obey. I like to do whatever I want. I pee in Clarice's living room." His eccentric mistress was not one to recoil at such behavior. Her dog, Lispector told an interviewer, "smokes cigarettes, drinks whisky and Coca-Cola. He's a bit neurotic." According to Benjamin Moser, author of a biography of Lispector, Ulisses had the habit of stealing cigarette butts from ashtrays and swallowing them, even when they were still smoldering. Guests to the house would be baffled when their half-smoked cigarettes seemed to disappear; Lispector, they said, would point out the guilty party, but would never punish him. She "calmly let him do whatever he wanted."

In another book for children, *The Woman Who Killed the Fish*, the author recalls how she acquired an earlier dog from a beggar woman in a Naples street. "One look at him was all I needed to fall in love with his face," she writes. Lispector took the mongrel home and fed him, and "he looked so happy to have me as his owner that he spent the whole day looking at me and wagging his tail." She christened him Dilermando, "a rather grand name which aptly described the wise expression and knowing ways which made him so endearing." When Clarice learned that the dog's original owner used to beat him, she vowed to treat her new pet with extra kindness. The love was mutual: "Dilermando liked me so much that he almost went crazy when he smelled with his snout my woman-mother scent and the scent of the perfume I always wear."

Lispector had been living in Naples for some years when her husband, a Brazilian diplomat, was posted to Switzerland and she had to face the heartbreak of leaving Dilermando behind (she'd been informed—inaccurately, as it turned out—that the Swiss didn't allow dogs in hotels). The separation was terribly painful, and for a long time Clarice couldn't stand to see a dog in the street. "I don't like to look at them," she wrote to her sisters. "You don't know what a revelation it was for me to have a dog, to see and feel the material a dog is made of. . . . Dilermando was something of my own that I didn't have to share with anybody else."

I know just how she feels. Although I'm never jealous of Grisby's affection for others, there's a way in which I want him to myself. The things we most like to do together are ordinary things—taking walks, exploring, driving to our dirty beach—

but they're things we enjoy in a private way, special to the two of us. Today, for example, I spent about twenty minutes blowing soap bubbles and watching Grisby try to bite them. Every time a bubble burst, he seemed confused but triumphant, as though he'd conquered an enemy without quite knowing how. If anyone else had been there, I'd have continued our game, but part of my attention would have been on the other person's presence in the room—his expectations, her needs, the conversation—making my focus on Grisby less singular. But when we're alone together, he and I share a special kind of closeness. At these moments, I can turn myself over to him, see things from his point of view, and let his way of doing things become my way.

As a kind of confession (and, perhaps, expiation) of her guilt at leaving Dilermando behind, Lispector wrote "The Crime of the Mathematics Professor," a story that was published in a Brazilian newspaper in 1945. In this tale, a man abandons his dog to travel abroad with his family. Everyone accepts his decision; no one blames him for leaving the dog behind; but many years later, the professor realizes the enormity of his crime, and the terrible suffering he so thoughtlessly inflicted on the animal. What's especially horrible about his transgression is that it can't be forgiven—not because it's such a dark sin but, on the contrary, because it's not a sin at all: a dog's life is irrelevant to God. The sinner, then, can't be consoled. "There are so many ways to be guilty and to be lost forever; I chose to wound a dog. Because I knew it wasn't much and I couldn't be punished for it," thinks the professor. "Only now do I understand that it really is exempt from punishment, and forever."

It's ironic that Clarice Lispector named her dog Ulisses (the Roman version of Odysseus), because although Homer's Odysseus is considered a great hero, he also cruelly abandons his dog—an act of callous disregard that, like the crime of the mathematics professor, is treated neither as a transgression nor as a character flaw. Yet even when we leave our dogs in good hands, as Lispector left Dilermando, they may still suffer deeply, sometimes even fatally. In *The Odyssey*, Homer informs us that Odysseus bred Argos himself, and as a young man, he used to take the dog out hunting ("there was not a wild beast in the forest that could get away from him when he was once on its tracks"). After his master abandons him, however, Argos is subsequently neglected by the women of the family and finally even by the servants, until twenty years later he is discovered, neglected and "full of fleas," lying "on the heaps of mule and cow dung that lay in front of the stable doors."

The much-vaunted "reunion" between master and hound is not, in reality, a reunion at all. Poignantly, Argos recognizes his disguised master after twenty years, and in joyful anticipation, "he dropped his ears and wagged his tail." But Odysseus is on the other side of the yard. Argos is unable to approach him, and Odysseus can't acknowledge the dog without giving himself away, although Homer informs us that Odysseus does shed a secret tear. Argos, after waiting twenty years to see his master again, dies after a single glimpse.

Ulysses was also the name of the Saint-Germain pointer belonging to the French journalist Roger Grenier, the author of *The Difficulty of Being a Dog*. According to French Kennel Club rules, purebred dogs must have names that begin with a certain letter according to their year of birth. When Grenier's dog was born, the year's letter was *U*. The author claims he dismissed the name Ulysses as too popular and banal and, after considering various alternatives (Ulric, Ursus, Uriel, Ugolin, Uranus), decided he wanted to call the dog Ubu, after Alfred Jarry's play *Ubu Roi*. It's unclear why the dog ended up being called Ulysses after all. Significantly, the French title of Grenier's memoir translates as "The Tears of Ulysses," so perhaps the author named his dog in honor of the ancient hero who weeps at the sight of his faithful old hound. Odysseus's suffering makes him heroic; Penelope is also heroic in her fidelity. Argos is just as faithful, but his ordeal has no payoff. No one considers Argos to be heroic. He's just doing what dogs do—waiting at home for his master.

Those twenty years must have seemed interminable to the neglected pet. Ironically, the more deeply we love a dog, the more quickly its life seems to pass. Grisby turned eight this year (according to *How to Raise and Train a French Bulldog*, the life span of the breed is between eight and twelve years). While he's still healthy and active—yesterday we walked about two miles together in the woods, then cooled off with a dip in the river—he definitely tires more easily and has the beginnings of osteoarthritis. I tire more easily, too, but I'm still not too old to be struck, at the sight of certain expressions on Grisby's face, by thunderbolts of adoration that make it impossible for me to resist bending down to kiss him on the face and snout, to caress his beautiful ears.

Dogs don't age the way people do. They may slow down and grow bad-tempered, but the physical changes aren't as always obvious. Humans grow small and wrinkled, but Grisby's always been that way. At eight, he's still cute, and I'm not the only one to think so. People often assume he's a puppy, and are surprised when they learn he's eight years old, not eight months. He hasn't grown into a scruffy hound, as some older dogs seem to do. Or perhaps he has; perhaps I'm simply blinded by love, like Elizabeth Barrett Browning, who, when she looked at the fourteen-year-old, overweight, aging Flush, saw the same endearing dog she knew and loved, with all his charming ways. But others had a different—perhaps a more objective—opinion. In one of her husband's poems, Flush appears as an "old dog, bald and blindish" that accompanies the poet on his evening walks, and one of Elizabeth's friends in Venice recalled Flush as "an old mangy creature, an uncomfortable fellow-passenger in a vettura."

In *Behind the Bathtub*, Colette Audry remarks how strange it seems to see chic, perfectly accessorized Parisian women accompanied by senile, incontinent old dogs—yet, she adds, this isn't by any means an uncommon sight. We keep our dogs to the very end, she observes, even when they are "mange-ridden, scabbed with eczema, half-paralyzed, cataracts in both eyes—not to mention that dreadful old dog smell." We continue to love our dogs, Audry explains, even when they're old wrecks, "slobbering, overweight, rheumy-eyed," because our love for them isn't contingent, as it often is with human partners, on physical well-being and strength of mind. Dogs adore us steadily and without change until the end, which makes it easy to love them back the same way. The "slobbering mongrels you see trailing along the sidewalk, with their bandy legs and raw rumps," observes Audry, "are by no means the least-loved members of the canine community."

Nonetheless, even animal lovers have a hard time appreciating that in some cases the loss of a pet can be as painful as the loss of a parent or child. "Why get yourself in a state?" people say. "After all, it's only a dog. You can get another." How such comments must hurt! You're mocked if you give your dog a lavish send-off (like Billie Holiday, who had her poodle cremated in her best mink coat). Psychologists suggest that it's normal to grieve deeply for a week after the loss of a beloved dog, and to feel a lingering sadness for a month or two more. After that, you should pull yourself together and move on. If you can't, your grief is considered excessive, inappropriate, pathological, or misdirected—in other words, you're "really" mourning for something else.

These days, I often catch myself wondering how much longer I've got with Grisby. I see danger everywhere for a small bulldog: careless drivers, reckless pedestrians, discarded needles, toxic trash (to imagine such disasters is, I realize, a superstitious way of warding them off). I try to be sure Grisby gets plenty of exercise, but am I keeping him fit or wearing him out? He loves to go out in the sun, but will he get overheated? If I inadvertently cause his death, could I ever forgive myself? Would he haunt me from beyond the grave? Will I hear the ghostly sound of his jingling collar, hear his nails clicking on the wood floor in the dead of night?

Will I recover from Grisby's death in time to own another dog, or will I be too old? "When you love a dog and it loves you, the lack of synchronization between human and animal life is bound to bring sorrow," writes Roger Grenier, who was in his sixties when he lost his Ulysses—the same age as Princess Marie Bonaparte when she lost Topsy, her famous chow. "A dog's life is so short," observed the princess; "to have one, to love one, is, if one is still young enough, gratuitously to invite Death into one's house." Yet Grenier describes receiving a phone call from the author and comedienne Madame Simone to inform him that her dog had just died, and to inquire where she might get another. "She was 95 at the time. What optimism!" remarks Grenier. "Perhaps she was right, since she lived to be 107, some say 110. So she still had just about the duration of a canine existence before her." I like this method of calculation. By this measurement, I have at least three dogs ahead of me.

V E N O M

VENOM WAS A controversial bulldog owned by the eccentric Amelia (Emily) Stewart, the wife of Lord Castlereagh, the British foreign secretary and the leader of the House of Commons from 1812 to 1822. The daughter of an earl, Lady Castlereagh was renowned for her beauty, her candid opinions, and her excessive love of animals. She kept a private menagerie at the couple's country residence, Loring Hall in Kent; her contemporary Grantley Berkeley speculated that many of Lady Castlereagh's exotic creatures had been sent to her as gifts from her husband's subordinates in the War Office who hoped to ingratiate themselves with him. "A tiger perhaps helped one to an embassy," reflected Berkeley, "and an armadillo made another at least a Secretary of Legation." Regardless of their origin, the beasts certainly made a powerful impression. "Returned to town, having first seen all Lady Castlereagh's establishment for animals," wrote the social observer Mrs. Arbuthnot in her diary. "She has got an antelope, Kangaroos, Emeus, Ostriches and a tiger, which Ld Combermere brought from the West Indies for the Duke of Wellington and which the Duke gave to Lady C. It seemed very vicious & growled at us."

Everyone knew Lord Castlereagh was devoted to his wife; his letters to her, written from abroad, are tender, endearing, and solicitous. He was apparently happy to indulge her unusual whims, which included, according to those who knew her, "a peculiar and not a praiseworthy partiality for large mastiffs," and her husband's "entire good nature and passive-ness to her pursuits induced him to bear with these savage companions." In 1805, an American visitor to the couple's home at 18 St. James's Square recalled two of Emily's bulldogs lying in front of the fireplace. ("Contradicting their looks," he observed, "they proved good-natured.") When he visited Castlereagh's country residence, Loring Hall, the same visitor gave a little more information: "Two pet dogs had the run of the rooms, *Venom* and *Fury*—in name only, not conduct." While Fury is seldom mentioned again, Venom, a female, became Lady Castlereagh's special pet, rarely leaving her side. When Emily accompanied her husband to the second Congress of Vienna in 1814, Venom went with them. A noblewoman, Countess Brownlow, described the ordeal of having to share a carriage with the Castlereaghs. Lord Castlereagh was more than six feet tall and took up a great deal of room; even worse, complains the countess, "sharing the small space with us was *Lady Castlereagh's fat bulldog*, poor dear Venom—she was a great pet, but her fat and heavy body was not exactly comfortable on one's feet."

Venom caused two minor controversies. The first occurred in August 1817, when Castlereagh was bitten and his hand severely lacerated by one of his wife's dogs. According to press reports, the injury was a serious one, with "the sinews of the first and second fingers being separated, and the nail and top of the first finger being nearly torn off." Castlereagh's father was concerned about him, but restrained himself from writing "lest my ill will and discontent might have burst forth unguardedly in terms somewhat too strong against Lady Castlereagh's favorite pets." The culprit was not named; however, one report referred to the guilty party as "a favorite dog of his lady's," and a French source, the Comtesse de Boigne, described the nameless offender as a "bulldog" that Lady Castlereagh "had hitherto overwhelmed with attentions and caresses," which makes Venom the obvious suspect. ("It was not until four months later, when Lord Castlereagh was completely cured," wrote the comtesse, "that she on her own initiative got rid of the dog.") In 1817, an opponent of Castlereagh's published a satirical pamphlet describing a mock trial in which a dog, standing accused of biting Lord Castlereagh, is unanimously acquitted. Such was Castlereagh's unpopularity, claimed the satirist, that no juror would convict the dog.

The second scandal was caused by the fact that Lady Castlereagh was frequently observed, according to one source, "driving in an open carriage in the Park, with a full-sized *Bulldog* occupying the place in the carriage which is now usually held by a Scotch Terrier or a King Charles Spaniel." What made this behavior particularly controversial was that Lady Castlereagh was a fashionable member of Regency society and an arbiter of social respectability; it would have been surprising to see a lady in her position riding alone in an open carriage, and even more inappropriate for her to be accompanied by a large bulldog named Venom. In Regency England, bulldogs were not the lovable pets of today, but hefty, intimidating beasts (which may explain why Lady Castlereagh's bulldogs are sometimes mistakenly referred to as mastiffs), cursed by the stigma of the bullring and linked indelibly to the lower classes. They were rarely kept as pets, and their ferocious appearance was generally considered brutish and lower class. William Youatt, author of a popular nineteenth-century dog lovers' manual, asserted: "The bull-dog is scarcely capable of any education, and is fitted for nothing but ferocity and combat." He also warned that any young man who acquired a bulldog would "speedily become profligate and debased."

"Who will believe that animals closely resembling the Italian greyhound, the bloodhound, the bull-dog or Blenheim spaniel, etc.—so unlike wild Canidae—ever existed freely in a state of nature?" wonders Darwin in the opening pages of *On the Origin of Species*. Behind every dog breed is a cultural history and ethnography, as well as a story of genetic descent. "Everyone I know likes stories about the origin of dogs," writes Donna Haraway in *The Companion Species Manifesto*. "Overstuffed with significance for their avid consumers, these stories are the stuff of high romance and sober science all mixed up together. Histories of human migrations and exchanges, the nature of technology, the meaning of wildness, and the relations of colonizers and colonized suffuse these stories." The bulldog story is not so romantic, though, like the stories behind most dog breeds, it's closely linked to issues of power, nationality, and class.

In the Middle Ages, the English nobility believed that beef had a superior flavor if bulls were made to exercise before being slaughtered. As a result, it became common practice to bait them with dogs, usually mastiffs—a barbaric practice that was believed to thin the bulls' blood and make their flesh tender for the slaughter (bulls as well as cows are used for beef). This tradition became so entrenched that many parts of England had laws actually requiring that bulls be baited before they were slaughtered. However, bullbaiting, like bearbaiting, was also popular in the late eighteenth century and early nineteenth century as a "sport"; baiting sessions were widely advertised, and locals could place bets on their outcome.

Mastiffs—the original *pugnaces britanniae* (see BULL'S-EYE)—
proved too large and slow for the task, so breeders developed new
strains by crossing the mastiff with smaller dogs such as the pug,
which had been brought to England in the sixteenth century.
From these experiments in crossbreeding came the bullmastiff,
the boxer, and, most successful of all for this grisly task, the English bulldog, selected for its strength, muscularity, fortitude,
and stubborn ability to cling on at all costs. As Christopher Smart
wrote in 1722, "Of all the dogs, it stands confessed, / Your English Bulldogs are the best."

Rings built for bullbaiting were circular stone buildings
rather like Roman amphitheaters, with tiered seating for spectators, who could place bets on which dog would bring down the
bull. This poor creature would have a collar around its neck attached to a rope about twenty yards long fastened to an iron ring
in the ground. One by one, each dog would be loosed, and each
would be given the same amount of time to bait the bull, which
would grow weaker with each assault. The dog would creep up
on its belly so the bull couldn't toss it up in the air, although this
often happened anyway, much to the thrill of the crowd. In fact,
a famous bulldog breeder named Caleb Baldwin enjoyed widespread fame for his ability to break his dog's fall whenever it was
tossed by rushing in and catching it in his arms.

To win the fight, the bulldog had to fasten itself onto the bull's
nose, its most tender part, and cling on no matter how much the
bull tried to shake it off. For this purpose, bulldogs were bred to
have wide mouths, powerful jaws, short snouts that allowed them
to breathe while being tossed from side to side, and wrinkled
foreheads to channel blood away from their eyes. When the bull
was finally overcome, it would sink to the ground and remained
pinned there by the dog, while the butcher slipped into the pen
to finish it off with a blow to the head.

According to *The Illustrated London Reading Book*, published in 1851: "Of all dogs, none surpass in obstinacy and ferocity the Bull-dog. The head is broad and thick, the lower jaw generally projects so that the under teeth advance beyond the upper, the eyes are scowling, and the whole expression calculated to inspire terror." When bull- and bearbaiting were both banned in 1835, it seemed for a while the bulldog might become extinct. Baiting was replaced with dogfighting (see BULL'S-EYE), for which the English bulldogs had the necessary aggressiveness but not the agility. New, more supple and equally tenacious strains were developed by means of crossing the English bulldog with various terriers, leading to the success of breeds like the pit bull and bull terrier. Today's version of the English bulldog is a modern dog, the result of further crossbreeding with the pug. These square, stout dogs, though excellent pets and loyal companions, would never have been able to stand the rigors of the bullring.

Naturally sedate, the modern English bulldog is prone to weight gain, and soon grows too heavy to run. Weight can be an issue for the French bulldog, too, especially in later life, when extra pounds can cause extra stress on the joints. I've seen French bulldogs that weigh as little as ten or twelve pounds, but Grisby has always been a big boy, and for the last three or four years he's been a few pounds overweight. Our regular vet has never brought up the subject, but the vet who issued his health certificate for a trip to England was very critical. "He's overweight," she said flatly, as I lowered a sheepish Grisby onto the scales. "He must lose five pounds." She immediately switched him to a costly organic dog food that, she said, would help him lose weight, though in fact it only made him unusually flatulent. "His coat could be softer, too," she said coldly, running her fingers disparagingly through his fur. (Perhaps it was not an error when the vet's technician, on his paperwork, described him as "pie ball.")

Grisby looked hurt, and as I left the vet's office with a sack of "balanced nutritional formula" (on which she's probably getting a kickback), I felt insulted myself. Grisby leads an active life and gets plenty of exercise. It's true that he did develop a taste for human food at an impressionable age, and though he's not allowed to beg at the table and has to sit for them, he does enjoy his daily treats (what dog doesn't?). I can't help identifying with him in this. I, too, have a fatal sweet tooth, a weakness for creamy desserts. I'd rather go hungry than face a plate of steamed vegetables; there's nothing that depresses me more than the thought of a "nutritionally balanced diet." (My hair could be thicker, too.)

I shouldn't have taken the vet's comments to heart—she was criticizing Grisby's weight, after all, not mine—but her offhandedness struck me as tactless. While Grisby may not have understood her words, surely, as an experienced vet, she must appreciate how deeply owners identify with their pets, and how blind they can be to the way others see them. Any criticism of Grisby is a criticism of me, and her comments cut me to the quick. His coat could be softer? Is she blind? His coat is beautiful! Overweight? That's just plain rude. She herself wasn't exactly a supermodel. Anyway, Grisby is perfect, a big, muscular bulldog with a good healthy appetite. Seriously, why make a big fuss about two or three extra pounds?

Yesterday, to cool off in the middle of a hot summer afternoon, David, Grisby, and I went to get frozen custard. After preparing our order, the man behind the counter asked us to wait for a moment, then handed us an extra treat—a small, bulldog-size cone he'd prepared especially for my pet. Clearly, this didn't fit the vet's dietary regimen; should I then have refused it, right in front of Grisby's eyes? If so, I need to ask: What's the point of having a dog at all?

WESSEX

MOST OF THOMAS HARDY'S novels are set in a fictional English county called Wessex, a thinly disguised portrait of the author's native Dorset. Wessex was also the name that Hardy's second wife, Florence, gave to the wirehaired terrier she acquired in 1913 to keep her company while her husband was away. Although he'd originally agreed to his wife's proposal of acquiring a dog, Hardy was deeply irritated when the terrier arrived. The cook, Mrs. Stanley, overheard him telling his wife: "Two things that you have brought into this house that I dislike are the dog Wessex and Mrs. Stanley's child." There's no evidence Hardy ever changed his mind about Mrs. Stanley's child, but Wessex quickly won him over, and soon Florence was writing proudly, of the dog: "My husband actually kisses him every night before he is carried off to bed."

Wessex, white with brown ears, was no ordinary dog—he was apparently descended from Edward VII's Caesar, a disreputable fox terrier that led the procession at his master's funeral sporting a Fabergé collar (see ORTIPO) studded with precious gems. Some say it was this inbred aristocratic lineage that caused the Hardys' dog to display such a foul temper; others blamed the dog's doting owners for poor training and overindulgence. Whatever the reason, Wessex's petulance was legendary. He was a snappy, aggressive, attention-hungry little creature, and the devotion he inspired in his owners was difficult to understand (one visitor referred to him as "Florence's unspeakable dog"). Like many dogs, Wessex wanted to defend his territory, and when strangers arrived at the Hardys' home, Max Gate, he'd go for their legs, often ripping their trousers. This is how he greeted the novelist John Galsworthy and the famous surgeon Sir Frederick Treves. Even when people were familiar to Wessex, he developed unpredictable hostility. The Hardys' postman kicked out two of Wessex's teeth in self-defense.

Part of the problem, it seems, was simply that that the pooch was spoiled rotten. In her diary, Lady Cynthia Asquith described him as "the most despotic dog guests have ever suffered under." After a visit to the Hardys' home with the author J. M. Barrie, she complained, "Wessex was especially uninhibited at dinner time, most of which he spent not under, but *on* the table [see JIP], walking about unchecked, and contesting every single forkful of food on its way from my plate to my mouth." Hardy was a hopelessly indulgent master, taking the dog for jaunts in the car, buying him a feather comforter to nap on, and driving him into a frenzy of delight by feeding him on cheese rind and other treats. On Christmas Day, Wessex was allowed to eat goose and plum pudding until he threw up (the mess was left to the maid).

Florence also spoiled the dog, but in a less obvious way. Like a doting mother unable to acknowledge her son's delinquency, she refused to blame Wessex for his destructive behavior, finding one excuse after another for his infractions. In a letter to her friend Rebekah Owen, who must have experienced the dog's hostility, she wrote: "Wessie sends his love. He is really a good dog—but all of his kind are fighters—more or less."

According to the author J. M. Barrie, a company that manufactured radio sets sent one to Hardy as a gift. The device delighted the author—not for his own sake but (to the manufacturer's disappointment) because Wessex found it so entertaining. Listening to the set soon became the dog's favorite activity, and Hardy was in the habit of getting up early, going downstairs in his dressing gown, turning on the radio for Wessex, then going back to bed. The editor Sir John Squire, a friend of Hardy's, recalled an evening he spent at Max Gate. When Hardy got up to see his guest to the door, Wessex leaped up and began tugging at his master's trouser leg. "He won't let me leave the room until he's had a few minutes of the wireless," Hardy explained, turning on the radio. The dog would apparently sit on his haunches, gazing at the set, getting great pleasure from the broadcasts, and even had his favorite shows, though, according to his master, he preferred music to news and talks. After an outing with Hardy and Wessex, J. M. Barrie observed that the dog behaved perfectly well until five o'clock, when he began to whine and kick up a fuss. Hardy explained apologetically that he had to go home; it was time for *Children's Hour*, the dog's favorite show.

Despite his unruliness, Wessex was, as Hardy put it in the title of a poem about the dog, "A Popular Personage at Home." The irritable terrier had plenty of friends as well as enemies, and when he liked someone, he seemed to feel a preternatural bond. His special companions included T. E. Lawrence ("Lawrence of Arabia") and J. M. Barrie, whom, when they arrived at Max Gate, he would greet with a happy bark and wagging tail. One of his favorites was a friend of Thomas Hardy's named Henry Watkins, whom he welcomed one day with whines and whimpers instead of his usual joyful enthusiasm. He sat by Watkins and continued to display signs of distress all evening, touching him softly with his paw from time to time. Watkins left around ten o'clock. Early next morning, the telephone rang, and Wessex didn't bark at the sound, as he usually did. The call was from Watkins's son, who told Hardy his father had died quite suddenly about an hour after leaving Max Gate.

Wessex himself died at the age of thirteen, on December 27, 1926, after a short illness. In a letter to friends, Hardy wrote: "We miss him greatly, but he was in such misery with swelling and paralysis that it was a relief when a kind breath of chloroform administered in his sleep by 2 good-natured Doctors (not vets) made his sleep an endless one—A dog of such strong character required human doctors!" In his diary that night, he wrote, "Wessex sleeps outside the house for the first time in thirteen years." The terrier was buried in the shrubbery on the west side of Max Gate, where the author's earlier pets had their graves. His headstone, which still stands, reads: "THE FAMOUS DOG WESSEX, August 1913–27 Dec 1926. Faithful. Unflinching."

Florence Hardy was heartbroken. "Of course he was merely a dog, and not a good dog always," she confessed in a letter to her friend Sydney Cockerell, "but *thousands* (actually thousands) of afternoons and evenings I would have been alone but for him, and had always him to speak to." Then, realizing she's heading into dangerous territory, Florence stops herself. "But I mustn't write about him," she continues, "and I hope no one will ask me about him or mention his name." No doubt she was afraid of breaking down in tears. From loving her pet unconditionally for thirteen years, she must have learned that women who display deep feelings for their dogs—especially their unruly dogs—are dismissed as gushy and sentimental. They're believed to be missing something in their lives and using their pet as a substitute.

Perhaps they're not getting enough sex or desperately want to have children. "Excessive dog love makes people feel uncomfortable," writes Adam Gopnik. "It isn't the misdirection so much as the inequality, the disequilibrium between the complex intensity of human love and the pragmatism of animal acceptance."

Gopnik assumes, as most people do, that seeking affection from a dog is a sign of failure, that dog love is a surrogate for real love—human love, neither instrumental nor motivated by the drive for rewards. Here, Gopnik is in accord with cultural theorist Jean Baudrillard, who argues that "the pathos-laden presence of a dog, a cat, a tortoise or a canary is a testimonial to a failure of the interhuman relationship." Those who devote themselves to their dogs, according to popular opinion, often have trouble handling the problems and complexities of adult human interactions. They've retreated from real life, and regressed to the infantile world of pets. They're emotionally entrapped, unable to build normal social relationships or seek therapeutic help. Implicitly, such people are viewed as immature, incapable of functioning in the adult world. Yet is there really such a difference between "the complex intensity of human love" and "the pragmatism of animal acceptance"? Isn't all social behavior or connection underpinned and reinforced by strong emotion? Don't human babies learn to snuggle in exchange for their mothers' warmth and milk, just as puppies do? In her book *Dog Love*, Marjorie Garber suggests we should detach the notion of "substitute" from its assumed inferiority to a "real thing." "Don't all loves function, in a sense," she asks, "within a chain of substitutions?" Can't we learn to see dog love as different but equal?

The anthropologist Edmund Leach has a different explanation for why people feel anxious and uncomfortable about dog love. He argues that, more than other domestic animals, the dog has a highly ambiguous place in human life, standing at the threshold between savagery and civilization. Dogs are both within and beyond our control; they sleep in our beds but shit in the street. We call them our "best friends," yet at any moment we can have them "put down." The ways in which we categorize nonhuman animals (as fauna, as meat, as household pets) make sense only as long as these categories don't overlap. Most dog owners try to curtail their animals' more aggressive instincts by training them to obey the household rules. When the animals won't submit, we no longer know where we stand. Let your dog walk around on top of your dinner table and bite your friends, and people are going to find it odd.

The Hardys' inexplicable devotion to their naughty dog inevitably became the subject of gossip. What made it worse was that the couple talked about little else. Nobody wants to hear someone going on and on about her dog, not even Virginia Woolf, though she loved her own dog to distraction and wrote a book about a famous spaniel (see FLUSH). After paying the Hardys a visit in July 1926, five months before Wessex died, Woolf noted in her diary that she spent most of the visit listening to Florence nattering on about her dog, "who is evidently the real centre of her thoughts." Finally, Woolf managed to get a few words with Thomas Hardy. She was "beset with desire to hear him say something about his books," but "the dog kept cropping up. How he bit, how the inspector came out; how he was ill; and they could do nothing for him."

The love we have for our dogs is often inexplicable to other people, especially when our dogs are conspicuously ill-natured, like Wessex. It's also true that just because we find our own dogs fascinating doesn't mean we're always going to be interested in other people's. Reading about dogs is a different matter; in literature, they remain at a distance, veiled in layers of human language and projection; in the flesh, they're hairy strangers. I have to confess: I'm not a "dog person" (whatever that means); I'm a Grisby person. I have no special interest in other people's dogs;

I'm more interested in what those dogs tell me about the person. Occasionally I meet a dog I feel drawn to right from the start, but most of the time I felt nervous and uneasy around dogs, with their unfamiliar smells and unpredictable behavior. I suspect they can sense my unease and resent me for it, too.

This makes sense. I have no relationship with other people's dogs, just as other people have no relationship with Grisby. I certainly don't expect even my closest friends to see in him what I see. Still, when someone notices him sitting patiently beside me and says, "My dog would never do that," I can't help feeling a secret thrill, as though others have been given a glimpse of the innate nobility that, to me, is obvious in his poised appearance. Just by looking at him, I think, you can tell how amiable he is, how willing to please. You can see by the sweet way he trots along by my side that he's well adjusted, not nervous, irritable, or demanding. For this reason, the more closely other dogs resemble him, the more they interest me. If Grisby and I meet another bulldog or bull terrier at the park or on the street, we'll always stop for a moment. While Grisby engages his counterpart in a sniffing match, I'll chat with the other dog's owner, asking about the animal's weight, age, and disposition. We'll smile, nod, and pretend to admire, then walk away, both complacent, no doubt, in the knowledge that our own dog is unquestionably superior.

XOLOTL

THE MEXICAN ARTIST Frida Kahlo had plenty of pets, including monkeys, birds, and a tame fawn, but her self-confessed favorite was Señor Xolotl, a Mexican hairless dog. Endemic to Central America, these exotic and intelligent creatures date back to the time of the Aztecs; they're associated with the doglike deity Xolotl, the twin brother of Quetzalcoatl. In Aztec mythology, Xolotl travels to the underworld to retrieve the bones of the dead, which are used to create new life. He's also the god of sickness and physical deformity, and is often depicted in art as a dog-headed man with ragged ears and backward-facing feet. The ancient Aztecs regarded the Mexican hairless dog as malformed, owing to its bald skin, which is why they named it "the dog of Xolotl," or Xoloitzcuintle (Xolo for short).

For Kahlo, who was proud of her Mesoamerican heritage,

the Xolo had symbolic resonance. She also allied herself with
the animal's purported deformity since she had polio as a child,
and as an adult was often confined to a wheelchair as a result
of terrible injuries suffered in a traffic accident. Xolotl was not
only the god of the malformed; he was also the god of fire—
another reason why he appealed to Kahlo, who was known for
her passionate nature. The Xolo is—quite literally—a hot dog,
prized for the heat generated by its body, which, although no
different in temperature from the bodies of other dogs, feels hot
to the touch because of its baldness. The dog's warmth means
it makes a therapeutic sleeping companion for those who suffer
from constant aches and pains. Kahlo often referred to Señor
Xolotl as her "hot water bottle."

Surprisingly, the Xolo's importance in Aztec mythology didn't
prevent it from being used as meat. The dog was used as a con-
venient source of protein in pre-Hispanic Mexico, where Xolos
were raised in large numbers much as cattle are raised today,
with young dogs being castrated and fattened for market. The
flesh, considered a delicacy, was also consumed in ritual cere-mo-
nies since it was believed to cure various ailments. In addition,
people would often be buried with Xolos by their sides, freshly
sacrificed, to help guide them to the underworld. Predictably, the
breed became scarce, reaching a point of near extinction before
it was restored to prominence.

Señor Xolotl wasn't Kahlo's only Xolo—she actually had six
others, all of whom lived with Kahlo; her husband, Diego Rivera; and the rest of her menagerie in her home (Casa Azul in
Coyoacán, Mexico City). I've only ever had one dog at a time—as
a matter of fact, I've only ever had one dog—so it's difficult for
me to imagine having seven, one of whom is my special favorite. It seems a little unfair to all the others, like a mother who
loves one of her children more than the rest. On the other hand,
while most mothers claim not to have favorites, perhaps they of-

ten do (most children seem to think so), and maybe the instinct shouldn't be considered unnatural.

According to Kahlo, Señor Xolotl was the most beautiful and intelligent of her dogs, and barked the least, though he was also a rebel, like his mistress. He once urinated on a watercolor that Diego Rivera had just spread out on the floor to dry. Furious at the dog's disregard for his masterpiece, Rivera chased the animal around the house with a machete, but when Señor Xolotl was cornered, he simply wagged his tail as if enjoying the game, and Rivera had to forgive him. Another time, when Kahlo's lover Isamu Noguchi had to get dressed in a hurry to avoid running into Rivera, Señor Xolotl stole one of his socks. The delay led to an unpleasant encounter, with Rivera chasing Noguchi over the roof of Casa Azul; still, when all the fuss had died down, Señor Xolotl was permitted to keep the sock to chew on.

Clearly, there was a mythic dimension to Kahlo's relationship with her favorite dog. Not only did Señor Xolotl stand in for the rest of the pack; he also represented the breed itself and its association with the Aztec underworld. He appears in a number of Kahlo's paintings, including *Itzcuintli Dog with Me* (1938), *Self-Portrait with Dog and Monkey* (1945), and, most notably, *The Love Embrace of the Universe, the Earth (Mexico), Me, Diego and Señor Xolotl* (1949). In this painting, the artist, with her husband as a child in her arms, is embraced by both Cihuacoatl, the Aztec Earth Mother, and Xolotl, Lord of the Underworld.

Other female surrealists also identified themselves with symbolic creatures that appeared in their work. Leonora Carrington's totem animal was the horse; Remedios Varo had a special connection with the owl; Leonor Fini's talisman was the cat. The painter and collage artist Dorothea Tanning, fourth wife of Max Ernst, felt a kinship with the Pekingese. Her own dog, Katchina, appeared in many of her paintings, often in giant form; in two collages from 1967, Tanning replaced human heads with the head of Katchina. *Dorothea Tanning with Her Smart Dog* shows a smiling girl sitting with her dog-headed companion on a swing, and *Dorothea Tanning as a Dog* depicts Tanning herself with Katchina's head, reclining languidly on a couch.

Though not an exact equivalent, Xolotl is in some ways the counterpart of Anubis, the guardian of the Egyptian underworld. Always black, Anubis is sometimes depicted as a dog, sometimes as a jackal, and sometimes as a man with a dog's or jackal's head. To the ancient Egyptians, the dog was a guide to the hidden side of life—the place we enter in sleep and death. In the mythologies of many cultures, dogs are associated with the underworld, perhaps because they were seen roaming the battlefields, feeding on slain bodies. Their otherworldliness, moreover, relates to their double nature. Wild and domestic at the same time, they're guardians of the way between, boundary crossers between man and animal.

The best-known canine gatekeeper is Cerberus, the three-headed dog who, in Greek and Roman mythology, guards the entrance to Hades, preventing the ghosts of the dead from returning to earth. Other cultures have their own versions of the dog posted at the underworld gate. In *The Mabinogion*, the founding text of Welsh mythology and folklore, white spectral hounds called the Cŵn Annwn guard these doors. In Norse mythology, a bloodstained watchdog named Garmr stands at the gates of hell. Mythological dogs aren't always fierce and frightening. In most tarot decks, the dog appears as the companion of the Fool; in folklore, hounds make friends with jesters and idiots, and in fairy tales they often play the role of wise helpers, sometimes revealing themselves to be enchanted princes in disguise.

Deities like Anubis and Xolotl aren't dogs exactly; they're dog-headed men, or cynocephali. These creatures have been with us for a long time, and have appeared all over the world. Sightings of them go back to Greek antiquity, as David Gordon White explains in his fascinating book *Myths of the Dog-Man*. To some, the cynocephali are more than a myth; even today, people continue to meet them. In 1999, the *Fortean Times*, a British monthly magazine devoted to anomalous phenomena, published a letter from a woman who'd recently spotted a man with the head of a dog—"kind of like a basset hound with long floppy ears," she wrote. This letter prompted other readers to write in with their own accounts of dog-headed men, which seem to be encountered mainly in the north of England. "In Yorkshire folklore," wrote one correspondent, "they were called 'leather-heads,' due to the fact that from a distance they looked as if they were wearing leather flaps when it was actually their ears. They always caused a nuisance when they turned up. My grandfather used to tell me about them as a lad."

Dogs that look or behave in mysterious ways could turn out to be cynocephali, but they might also be domestic spirits in canine form. Señor Xolotl was so intelligent that some Mexican locals believed he was Kahlo's familiar—that is, the animal companion of a witch or sorcerer. The term was officially introduced into the language in the middle of the sixteenth century, when a series of new acts were passed in England imposing heavy penalties on anyone caught consulting, feeding, or rewarding an "evil wicked Spirit." However, magical familiars had already been around for centuries. According to Christina Hole's authoritative *Witchcraft in England*, in the Middle Ages, "the actual possession of any beast that might be supposed to be a familiar was a clear danger to anyone suspected of witchcraft, especially if he or she were known to treat it with affection."

In European folklore, familiars usually take the form of cats; even today, the neighborhood cat lady can be a scary figure, often believed—albeit unconsciously—to be a crone, hag, or witch. Other animals can also be familiars, though, including birds, mice, rats, hares, rabbits, owls, snakes, hedgehogs, ferrets, weasels, and toads. Although they're usually animals, familiars can also be imps or spirits that live in rings, lockets, or similar magical talismans (the philosopher Apollonius of Tyana wore a ring that was said to house his familiar); they can also emerge in the form of a genie from a lamp or bottle. Zulu familiars take the shape of resurrected corpses that have been magically reanimated—in other words, zombies.

Familiars arrive when the witch or sorcerer sends out a psychic call to summon the right spirit or animal, since familiars are sensitive to psychic vibrations. Sometimes, however, spirits are attracted to people with no interest in them, and these pesky imps will annoy their unwitting "owners" until they grow tired of the game and either leave of their own accord or are exorcised. In most cases, the familiar is sent out, usually at night, to perform the owner's errands, which may include acts that are beneficial as well as destructive. In most cultures, familiars are believed to remain faithful until their owner dies, at which point they either disappear or pass along to the owner's ancestors. They can exist independently of human beings, but if the familiar is wounded or killed, the same thing might happen to the owner, which is why it can be a risky business to send out your familiar to do battle on your behalf.

Owing to their natural loyalty, dogs were reputed to make excellent familiars. The German physician and occultist Cornelius Agrippa had a familiar in the form of a black poodle named Monsieur, who allegedly committed suicide by drowning himself in the River Saône after the death of his master. One of Agrippa's followers was Georg Sabel of the Rhineland, an occultist who called himself Dr. Johannes Faustus, and was apparently followed by a black poodle who could transform himself into a servant (see MATHE). When the alchemist and swindler Marco Bragadino was tortured and beheaded in Munich in 1591 his two black dogs were shot because they were believed to be "fiendish servants in the form of beasts." The thirteenth-century Irish sorceress Alice Kyteler was believed to have a dog familiar by the name of Robert or Robin Artisson, a "lesser demon" who was also her lover. The seventeenth-century Royalist general Prince Rupert of the Rhine was never seen without his large white poodle Boye, long rumored to be his familiar. Boye accompanied his master into battle during the English Civil War, and—at least according to Puritan propaganda—possessed supernatural powers. One source suggests the fortunes of war finally turned in the Roundheads' favor when Boye was deliberately shot and killed by a silver bullet in 1644, during the Battle of Marston Moor.

I'd like to think of Grisby as my familiar, but opinions are mixed on whether familiars can be pets. Some believe that, to acknowledge their subservient role, familiars should never be pampered but always treated as lesser beings. According to historian Emma Wilby, however, familiars "were often given down-to-earth, and frequently affectionate, nicknames" (she mentions an endearing pair named Grizell and Gridigut), which suggests their relationship with their masters was not unlike that between humans and their pets.

Of course, in literal terms, I acquired Grisby through PuppyFind .com, but this doesn't preclude the possibility that I unconsciously sent out psychic vibrations summoning an animal spirit. Did Grisby answer my call, knowing we'd make a perfect match? Or did I conjure him up myself, creating a living embodiment of my deepest desires? According to one source, a spirit that appears in the form of a bulldog is always accursed or accusatory, but anyone who believes this nonsense has obviously never met Grisby.

To work successfully as magical collaborators, familiars need to be closely attuned to their human partners, which is certainly true of Grisby and me. While familiars don't have to resemble their owners, Grisby and I do have a certain physical similarity, though this is impressionistic rather than obvious (and I may be the only one who sees it, or who finds it flattering). We're both broad-shouldered and big-boned; we both have wide mouths and large smiles, and while neither of us would be described as a classic beauty, we both have our admirers. Indeed, according to an article by Stanley Coren in *Psychology Today*, there are data suggesting people do prefer dogs that look similar to themselves—at least in terms of their facial characteristics, the shapes of their heads, and the way their hair hangs. We may look alike, but Grisby has a better temperament than I do and is more sociable; as a result, he has a lot more friends than I do.

I'd never describe Grisby as a shaman (though I sometimes think of him—in terms of the delightful misprint I found on a website translated from Russian—as a "French bullgod"), which is how the physician Renaldo Fischer describes his English bull-dog, Faccia Bello. In his book *The Shaman Bulldog: A Love Story*, Fischer recounts how Faccia Bello became his comfort and spiritual guide after the collapse of his marriage, and helped him to understand his connection with the natural world. While I was touched by the suffering and death of Faccia Bello, Fischer's New Age mindfulness grated on my nerves (he calls Faccia Bello "a wise animal, an elder who in his own clan would be very distinguished and honored"). *The Shaman Bulldog* made me realize that, in most cases, a dog is just a dog. Thankfully, Grisby is neither shaman nor guru nor archetype, but himself—my living, snorting companion. That's magic enough for me.

YOFI

SIGMUND FREUD FELT that dogs, unlike humans, display no ambivalence in their emotions. "One can love an animal," he once said to his patient the author H.D., "with such an extraordinary intensity: affection without ambivalence, the simplicity of a life free from the almost unbearable conflicts of civilization." While there was no ambivalence about Freud's love for his chow Yofi, there was a revealing ambiguity about her name, which is spelled differently in different sources. The word *yofi* means "beauty" in Hebrew, but writing in German, where the letter *y* is rarely used, Freud spelled it with a *j*. Other spellings of the dog's name found in notes and correspondence include "Jo-Fi," "Jo-Fie," "Jofie," and "Yofy." The pronunciation, however, was always the same: "yoffi."

Freud didn't discover he was a dog lover until he was in his seventies, and he did so after falling in love with Wolf, the black German shepherd acquired for his daughter Anna, in 1925, to protect her on her long solitary walks. As it turned out, Wolf became not only a protector but also a guide, a go-between, a companion, and a lot more besides. Roy R. Grinker, an American psychiatrist who was analyzed by Freud in Vienna in 1933, remembers how Wolf would start to bark as soon as the doorbell rang; when he entered the house, the dog would follow him into the waiting room and immediately start sniffing his genitals. As a result, says Grinker, he always entered Freud's office "with a high level of castration anxiety." When he attended a seminar held at Freud's home, Grinker recalls, he'd only just taken a seat when Wolf approached him with an ominous growl, prompting Anna Freud to remark that, although when he was younger he used to eviscerate sheep, Wolf was perfectly safe, and if he continued to bark, Grinker should simply pull his tail.

In 1928, three years after the arrival of Wolf, Freud finally acquired a dog of his own, a chow named Lün-Yu (again, the name is spelled differently in different sources), who was given to him by Anna's companion Dorothy Burlingham. Although they may look like cuddly little bears, chows can in fact be bad-tempered, aggressive, and territorial; owning one can actually raise the cost of homeowner's insurance, as some companies regard chows as "high-risk dogs." Sadly, at the age of fifteen months, Lün-Yu wandered off at a Salzburg railway station, and was later found dead on the tracks. Freud was reportedly heartbroken.

His grief was assuaged seven months later when Burlingham presented him with Lün-Yu's sister (interestingly, all Freud's dogs were females, and none were spayed). This was Yofi, who stayed by his side for the next seven years. "She is a charming creature," he wrote of his new friend, "so interesting, in her feminine characteristics too, wild, impulsive, gentle, intelligent and yet not dependent as dogs often are." Yofi remained in Freud's office during analytic sessions, and Roy Grinker remembers her occasionally interrupting his analysis by getting up to scratch at the door. Whenever she did so, Freud would get up and let her out, observing that Yofi had lost interest in the conversation. This irritated Grinker, who felt that Freud often seemed more focused on his dog than his patient. Sometimes Yofi was even more disruptive. During one particular session, Grinker grew very emotional and suddenly raised his voice, causing Yofi to leap aggressively onto the couch and pin him down. Grinker, by his own report, lay quietly with eyes closed, as he'd been taught to do if attacked by a bear.

The poet H.D. recalls Yofi in fonder terms, perhaps because she was herself a lifelong dog lover. In her book *Tribute to Freud*, she describes her first visit to the analyst's consulting room in Vienna, where she went to discuss the possibility of treatment. During her interview, "a little lion-like creature came padding toward me— a lioness, as it happened. She had emerged from the inner sanctum or manifested from under or behind the couch; I bent down to greet this creature." Freud warned her that Yofi was not always friendly to strangers; nevertheless, the dog snuggled her snout into H.D.'s hand and nuzzled against her shoulder. Yofi's acceptance of her, felt H.D., was what persuaded Freud to accept her case.

Freud acquired his love of chows from another of his patients, Princess Marie Bonaparte, whose female chow was the subject of her light and tender book *Topsy: The Story of a Golden-Haired Chow*. In this book, Bonaparte explains that only when she learned that Topsy had cancer did she feel, all of a sudden, a "passionate affection" for a dog that until that point had been but a "graceful toy" to her. She took Topsy to the Curie Institute for a series of radiation treatments that proved successful, and as she wrote her book, Topsy sat once more at the princess's feet, "proudly erect on her front paws."

Freud's son Martin said that in his later years, his father often kept track of the time by paying attention to Yofi, who always knew when the analytic hour was up. Freud and Yofi would eat together every day; their meals were different, but Freud would often slip Yofi some of his lunch (the professor, when eating, often experienced pain due to his diseased jaw). He wrote to tell H.D. that on the morning of his eightieth birthday, Yofi had come into his bedroom to show him her love. "How does a little animal know," he asked her, "when a birthday comes around?"

By the following year, however, Yofi was suffering from ovarian cysts, and Freud, like Topsy, had developed cancer on the right side of his jaw. To Marie Bonaparte, he wrote, "I wish you could have seen what sympathy Yofi shows me during these hellish days, as if she understood everything." The dog had two cysts removed early in the year, and at first it seemed as though she, too, like Topsy, was going to make a full recovery. Unfortunately, Yofi died of a heart attack not long after her operation. "One cannot easily get over seven years of intimacy," wrote her grieving master.

Luckily, there was another chow waiting in the wings—a dog called Lün, which Dorothy Burlingham had given Freud sometime earlier as a companion for Yofi. When they'd first been introduced, the dogs' personalities immediately clashed, and Yofi, who was in heat at the time, attacked and bit her rival. After this incident, Lün was returned to the Burlingham household, where she spent the next four years, waiting like an heir to the throne for the death of the current ruler. The day after Yofi died, Lün took her place by Freud's side, where she remained until the end. In 1938, when the Freud family fled to London to escape the Nazis, Lün went with them. The journey was far from smooth. At their first stop, Marie Bonaparte's villa at Saint-Cloud, Lün got into a fight with Topsy; then, when they finally arrived in London, Lün was taken away to what Freud described as an "animal asylum" and put in quarantine for six months.

By this time, the professor's cancer had spread, and he was gravely ill; still, he managed to make regular visits to Lün (as a newspaper report put it, "nothing could have kept the great scientist away from his dog friend"), who was waiting out her period of quarantine in a kennel at Ladbroke Grove, only three miles from Freud's new home in Hampstead. "I have never seen such happiness and understanding in an animal's eyes," said the kennel's director, describing one of the doctor's visits to his chow. "Freud played with her, talked to her, using all sorts of little terms of endearment, for fully an hour." When her quarantine was over, Lün was reunited with her master in Hampstead, but her joy in Freud's company was short-lived. His cancer had become so advanced that his jaw gave off a putrid smell, and Lün avoided him, cowering in the corner of his room rather than sitting at his feet, as she had done before. A few months later, Freud was dead.

Although neither Yofi nor Lün is mentioned in Freud's case studies, dogs do appear from time to time, usually helping the author to sniff out a clue or come to an insight. His most famous patient, Anna O., under hypnosis, recalled her disgust at seeing her English governess let her dog lap water from a drinking glass, a recollection that helped Freud get to the root of her problems. In another paper, the analyst explains that he came upon his theory of sublimation after reading about a famous surgeon who, as a youth, chopped off dogs' tails. Animals of other kinds also turn up in the case studies; they play vital roles in the cases of Little Hans, the Wolf Man, and the Rat Man, for example, but these papers tend to emphasize the negative role of animals as symbols in the development of children's phobias (fears of animals, according to Freud, are among the earliest neuroses).

The love of animals is also a central concern of childhood. It's often been observed that during the first years of life, children don't seem to be able to make a clear distinction between humans and nonhumans, behaving socially toward pets, talking to them and treating them as though they are members of the family. In fact, children relate to imagined feelings in animals before they relate to other humans' feelings, which is presumably why so many children's books focus on animal rather than human protagonists. With his dogs, Freud seemed to share an intimate bond, as though among these fellow creatures, the great professor could regress to the prelinguistic understanding of early childhood.

The legacy of dog love continued in the Freud family long after the patriarch's death. For the rest of her life Anna Freud kept chows (each with the same name: Yofi). Freud's grandson Clement Freud, a writer, broadcaster, and politician, was well known for his appearance on British television in a series of dog food commercials (first for Minced Morsels, then for Chunky Meat) that took advantage of his hangdog looks. Clement Freud's elder brother, the painter Lucian Freud, was also a dog lover. "I am impressed by their lack of arrogance, their ready eagerness, their animal pragmatism," he remarked in an interview. Over the years, he charted the development of his whippet Pluto from birth to death. His final painting includes his last dog, Eli, who was another whippet and Pluto's great-grandniece.

Freud's love for Yofi made it acceptable for other psychoanalysts to have dogs in their consulting rooms. Dorothy Burlingham had a gray Bedlington terrier that would sometimes emerge from under the couch during analytic sessions. The German psychoanalyst Karen Horney was rarely seen without her cocker spaniel Butschi, and the French analyst Jacques Lacan said that his boxer, Justine, named after the novel by the Marquis de Sade, was the only living being who could see him as he really was, without the veil of projections that characterize all human relationships. As adults, the former patients of Melanie Klein, an analyst of children, had fond memories of her Pekingese, Nanki-Poo; the pioneering American psychoanalyst Harry Stack Sullivan kept five dogs in his consulting room, a mother and four pups, who would run in and out through a pet door during sessions.

For a few years, I had my own therapy practice one evening a week, when I saw patients at home. At first, anxious about being taken seriously, I kept Grisby out of the office; later, when I felt more comfortable, I started asking my patients if they'd mind having my dog in the room. No one ever said no; I like to think the patients were all dog lovers, but perhaps they didn't feel they had a choice—after all, none of Freud's patients were brave enough to ask him to put Yofi out, though many later said they found her presence distracting. Grisby, however, was always quiet and well behaved. Before long, he learned to recognize my patients, and would express pleasure when they arrived, offering himself up for a pat or a scratch before settling down at my feet for the duration of the session. Unlike Yofi, he was no timekeeper, and usually had to be woken when the hour was up.

Some therapists who bring their dogs into the treatment claim the animals have special skills—they can calm the anxious, comfort the depressed, and help contain a crisis (some of these skills are shared by the specially trained therapy dogs that are used in hospitals and nursing homes). While Grisby has never displayed any talents of this kind, his silent presence always encouraged what the author and dog lover Elizabeth von Arnim called *recueillement*, the regathering of oneself in peace and quietude. His presence may not have been any great help to my patients, but it always gave me tremendous comfort to sense his soft breathing, and feel his familiar warmth at my feet.

ZÉMIRE

ZÉMIRE AND AZOR is the title of a comic opera by the Belgian composer André Grétry, based on the story of Beauty and the Beast. First produced in 1771, the opera was a great hit first at the court of Louis XVI in Paris, then at the czar's palace in St. Petersburg. As a result of the opera's popularity, Zémire— after Grétry's heroine—became, for a time, a fashionable name among well-bred female dogs, a trend that may have been started by Catherine the Great, whose favorite greyhound was named Zémire. The dog, like the empress, lived to a ripe old age, and was lavishly commemorated after her death. A porcelain figure of the hound adorned the Grand Hall in the Winter Palace and a marble column was erected to mark her tomb, inscribed with an epitaph by the French ambassador Count Ségur.

Zémire is an appropriate name for Grétry's soprano heroine:

it's a version of the Hebrew name Zemira (sometimes spelled Zemirah), used for both men and women, meaning "song." In fact, Grétry was not the only composer to use the name. In the late eighteenth century, Francesco Bianchi wrote a three-act opera called *Zemira*; this is also the name of an overture written around the same time by the Afro-Brazilian composer José Maurício Nunes García. The Zémire who interests me most, however, is one who came even earlier—the spaniel who belonged to Madame Antoinette du Ligier de la Garde Deshoulières, a French poet and intellectual acclaimed as the "Tenth Muse" at the court of Louis XIV. Her literary distinction was unusual for a woman—her work was praised by eminent figures of the time, including Voltaire and Corneille. Also unusual was the talent of her dog.

Madame Deshoulières, perceiving Zémire to be exceptionally alert, taught her the meaning of a considerable number of words, along with the objects to which they corresponded. When her mistress requested something (a glove, a shoe, or a handkerchief, for example), as soon as she named the item, Zémire would run and fetch it. If, when the dog returned, her mistress had left the room, Zémire would refuse to relinquish her precious cargo; the only people she would allow to take it from her mouth were those who she could be sure would deliver it directly to Madame Deshoulières, such as her lady's maid.

Language recognition isn't unusual in dogs, though it's rare to find a canine with such advanced linguistic skills as Zémire. Some, however, are even smarter. A 2010 article in the magazine *Popular Science* describes Chaser, a border collie trained by psychologists to identify and retrieve 1,022 objects by name. Chaser can apparently even figure out what object his trainers want when he's never heard the word before. He demonstrates the ability to use "fast mapping," a skill usually found in young children, whereby a hypothesis is quickly formed about the meaning of an unknown sound. Chaser's abilities suggest that, while dogs may be unable to learn language the same way we do, they have an understanding that goes beyond the simple association of sounds and object, and are able to come up with their own assumptions about the meaning of unfamiliar words.

Language learning was only one part of Zémire's repertoire; she was also a convincing actress. When her mistress said, "Go to bed, Zémire," the spaniel would go and lie down prettily in her basket, pretending to snore. At the words, "Wake up Zémire, and make yourself look nice," the spaniel would jump out of her basket, stand on a footstool in front of a mirror, and gaze at herself in admiration, glancing around to make sure her tail was straight. Apparently, this display of feminine vanity wasn't entirely a performance—Zémire, it seems, was a naturally fastidious pooch. If she got dusty on her daily walk, when she returned, she'd go over to the hand basin and put out her front paws one at a time, demanding to be washed with perfumed soap. Still, when she visited her dog friends, she'd run and play with the pack, happy to get dirty for once and forget about her courtly manners.

The French literary establishment finally acknowledged Madame Deshoulières after she spent many years struggling for recognition. The greatest honor came with her election as a member of the Academy of the Ricovrati of Padua; shortly after this tribute, she was invited to attend a formal literary function at a private home, an occasion that was definitely not dog-friendly. For the first time, Madame Deshoulières left Zémire at home overnight. The dog remained in her mistress's bedroom with plenty of food and drink, but the following morning, she was discovered lying dead on the bed, her food untouched. Upon discovering the tragedy, Zémire's mistress reportedly declared she'd gladly give up all her honors and rewards to have her pet back by her side.

The story of Zémire's death may well be apocryphal or exaggerated, but dogs, like humans, have really been known to die of a broken heart. Whatever the truth of this particular tale, it functions as a warning not to place too much value on academic achievements. Madame Deshoulières was highly intelligent herself, and perhaps for this reason, she valued intelligence in her dog. Yet in the end, it was not Zémire's cleverness that mattered but her loyalty and devotion. Intellectual honors, however prestigious, mean nothing to a dog, and perhaps this is a sensible attitude. Grisby, as I've mentioned before, isn't the smartest hound in the pack. He recognizes his name and understands the commands "sit" and "stay" (which isn't to say he obeys them), but these achievements, when compared with those of dogs like Chaser and Zémire, are remarkably unimpressive. Still, if he were smarter, Grisby would be no happier, nor would we be more attached. In both humans and dogs, intelligence is overrated; most often, it leads not to a better quality of life but to neurosis, anxiety, and stress.

The Greek philosopher Diogenes said that humans would do well to study the dog. This is excellent advice, and Grisby is a perfect case in point. He never worries, for example, about where his next meal's coming from or where he's going to sleep—he'll eat almost anything and makes his bed wherever he happens to lie down. He has no shame about toilet functions—in fact, he evidently enjoys them. His lack of self-consciousness makes him a joy to watch. When he finds an old dog chew under the couch, his delight is palpable. He'll slobber over it for a while, give it a good chomp, then, crouching in a play bow, he'll toss it up into the air, pounce on it, catch it, and run across the room with it in his mouth, as if his wad of slimy rawhide were a prize that everyone wants to steal.

To Plato, this capacity to find delight in small pleasures makes the canine "the most philosophical" of all beasts. In his prologue to *Gargantua and Pantagruel*, the French author Rabelais chews on Plato's epigram. "Did you ever see a dog with a marrowbone in his mouth?" he asks. "If you have seen him, you might have remarked with what devotion and circumspectness he wards and watcheth it: with what care he keeps it: how fervently he holds it: how prudently he gobbets it: with what affection he breaks it: and with what diligence he sucks it. To what end all this? What moveth him to take all these pains? What are the hopes of his labour? What doth he expect to reap thereby? Nothing but a little marrow." Right now, as I write, I can hear Grisby on the floor at my feet, demonstrating the truth of this, chewing on his favorite rawhide bone. The noises are symphonic: grunting, slurping, snorting, snarling, smacking his lips.

As humans, we can't abandon ourselves so readily to our pleasures, mainly because we live in time. We always have one eye on the clock or calendar so we can plan our days, our weeks, and our years. Everything we do takes place within the context of an unknown future and a past that can't be changed. Still, we dwell on both, fretting and festering—Why did I say that? What did I miss? What if I get sick? Will I be late? Why did I wear these shoes?

While she sits, making "signs on paper that no dog can read," Princess Marie Bonaparte observes that her chow Topsy would sit at her feet and "simply inhale the scented June air." This, she concludes, is why Topsy, "whose happiness is confined to the narrow limits of each day," is wiser than her mistress, who feels compelled to capture each moment in words. This capacity to inhabit the present is one of the things I admire most about Grisby. He always seems so delighted to be in the here and now, to engage in life itself. This is best seen in the way he falls asleep. He goes from waking to snoring immediately, anywhere, at any time. The period between waking and sleeping states is when we humans lie going over things in our heads, replaying the day's events, worrying about the future, about money, relationships, family. Did you ever know a dog with insomnia? This ability to go immediately to sleep is one of the many signs of Grisby's care-free nature. He's categorically untroubled. In the morning, he wakes, takes a stretch and a shake, and he's right in the middle of life again. He never wonders where we're going, or misses where we've been. For him, each moment stands alone, with no refer-ence to the moments before it or the ones to come. The richness of his existence is especially remarkable given the fact that he does almost the same thing every day.

Every morning, for example, our routine is the same: I take Grisby around the block for a walk. Yet every morning, when he realizes that—yes, we're going outside!—he runs to the door happily, looking up at me as if he can hardly believe his luck. He dances around in excitement while we're waiting for the elevator; when we get to the lobby downstairs, he rushes ahead, straining on his leash. Out in the street, he bounds along excitedly, snout in the air, ears alert. He crouches or cocks his leg to urinate, then takes a big, happy dump. As soon as he's finished, he kicks his back legs and dashes ahead with glee. He'll stop at the parking meters, cock his leg again, and sprinkle a few more drops on some deserv-ing post; then he'll trot ahead, casting a backward glance at me that seems to say, "Hey, isn't this great? Aren't we having fun?" If we meet another dog, he'll be overjoyed, bounding up for a shared sniff, as if to say, "Hello! Are you taking a walk? Guess what—we're taking a walk too!" I've always been a depressive type my-self, with a cynical disposition, but I can never be cynical around this bouncing bulldog. It's impossible to be in the presence of so much joy without being lifted up by it, if only a little.

Yet although he may be a god to me, Grisby is mortal, like the rest of us, and starting to show his age. He'll die, no doubt, in the next four or five years, a thought that makes me less unhappy when I remember how unaware he is of his own mortality. I re-mind myself to follow his example and live in the present instead of worrying about what life will be like without him.

This morning, we went to the park and he ran through the grass happily as usual, throwing me joyful glances from time to time. In Grisby's mind, things have always been this way and will go on this way forever, the two of us together, having adventures every day. And in a way—a real way—he's right. If it's true that most of our feelings for our dogs are based on projection—if a dog's "personality" is largely produced by our own unconscious— then I have nothing to worry about. Every dog I ever own will be a Grisby.

POSTSCRIPT

ON JANUARY 25, 2014, about four months after I'd finished work on this book, Grisby died in the night at an animal hospital where he was being treated for pancreatitis (unrelated to his weight, I'd like to add). He lived for eight years and six months, and died a joyful dog in the prime of life, alert and playful, always happy, with little experience of pain or suffering. He was devoted and beloved.

THE GREAT GRISBY,
JULY 2005–JANUARY 2014.

Notes

CHAPTER 1: ATMA

5 *deeply attached to his dogs:* Information is taken from Cartwright, *Historical Dictionary of Schopenhauer's Philosophy*, 136.

5 *"To anyone who needs lively entertainment":* Schopenhauer, "Ideas Concerning the Intellect Generally," 82.

6 *Schopenhauer got up and moved his poodle's seat:* Anecdote is from Wallace and Anderson, *Life of Arthur Schopenhauer*, 174.

6 *"When I see how man":* Schopenhauer, *Studies in Pessimism*, 21.

6 *every one of them the same name:* Sometimes the name is given as Atman; *The Stanford Encyclopedia of Philosophy* has Atma (see Wicks), which is the version I've chosen to use.

7 *"on Mount Gertrude":* Benfer, "Gertrude and Alice."

7 *the rhythm of the dog's breathing:* Ibid.

7 *When Basket died in 1937:* Information about Basket and Basket II compiled from Stein, *The Autobiography of Alice B. Toklas*, 251–52; Souhami, *Gertrude and Alice*, 171; and Malcolm, *Two Lives*, 23, 137–38.

7 *"His going has stunned me":* Malcolm, *Two Lives*, 137–38.

9 *"You will likely call":* Pronek, *How to Raise and Train a French Bulldog*, 11.

9 *not only is his name Bashan:* Mann, *Bashan and I*, 97.

9 *fashions in dog names go in cycles:* Classical dog names compiled from Abbott, *Society & Politics in Ancient Rome*, 187–88; and Ovid, *Metamorphoses*: "Actaeon," Book 3, 138–41, 206–33, and "Cephalus and Procris," Book 7, 394–97, 753–93.

10 *15 percent of British dog owners:* ICM research. "Of the 1172 adult pet owners questioned, 15 per cent admitted that they preferred their

pet to their cousin and six per cent said that they preferred their pet to their partner"—www.icmresearch.com/media-centre/post/a-third-prefer-pets-to-family.

10 *Sixteen percent listed their dogs:* Ibid.

10 *the same names turn up in top-ten lists:* Dog name trends are analyzed each year by VPI Pet Insurance; according to 2012 records, the ten most popular dog names based on the policyholders' inquiries were (1) Max, (2) Bailey, (3) Buddy, (4) Molly, (5) Maggie, (6) Lucy, (7) Daisy, (8) Bella, (9) Jake, and (10) Rocky—www.petinsurance.com/healthzone/pet-articles/new-pets/How-We-Name-Our-Pets.aspx.

11 *Other common names:* Popular names for pit bulls are listed at a pit bull chat website, www.pitbull-chat.com/showthread.php/59463-What-are-the-top-10-most-used-names-for-the-American-Pit-Bull-Terrier.

11 *San Francisco Health Department records:* According to the San Francisco Health Department records, of about 375 dog bites recorded from 1994 to 1997, seven were perpetrated by a Rocky. Next were Mugsy, Max, and Zeke, each tied with six bites. Accessed 21 September 2013, http://www.nanceestar.DogNamesPage1/html.

11 *Norman Mailer—who once got into a street brawl:* Manso, *Mailer*, 221.

CHAPTER 2: BULL'S-EYE

13 *"he owns a bull terrier":* See http://en.wikipedia.org/wiki/Bill_Sikes.

13 *"a white shaggy dog":* Dickens, *Oliver Twist* (1866), 91.

13 *illustrators like George Cruikshank:* For Cruikshank's illustrations, see Dickens, *Oliver Twist* (1838), 216, 312.

13 *the modern, long-faced bull terrier:* Alexander, *Bull Terriers*, 13.

14 *"faults of temper":* Dickens, *Oliver Twist* (1866), 109.

14 *"downiest of the lot":* Ibid., 138.

14 *Sikes constantly denigrates his dog:* Ibid., 110, 144.

14 *treats his girlfriend:* Ibid., 381.

14 *"carry out new evidences":* Ibid., 383.

14 *hurls himself at the dead man's shoulders:* Ibid., 410.

14 *"striking his head":* Ibid., 411.

15 *"put his great paws":* Dickens, *Little Dorrit*, 203.

15 *"seized the dog with both hands"*: Ibid., 479.

15 *"deeply ashamed"*: Ibid., 480.

15 *"to the feet of his mistress"*: Ibid.

15 *"He was bad-tempered"*: Camus, *The Stranger*, 45.

15 *"great black and white"*: Brontë, *Jane Eyre*, 112.

15 *"a lion-like creature"*: Ibid., 107.

16 *"then he jumped up"*: Ibid., 436.

16 *"workmen are always ready"*: Audry, *Behind the Bathtub*, 53.

16 *"the manliest dog on the planet"*: See www.bluebeards-revenge.co.uk/ blog/english-bulldog-voted-manliest-dog-on-the-planet/.

16 *all kinds of macho objects and activities*: For bulldog-themed products, see http://en.wikipedia.org/wiki/Bulldog_%28disambiguation%29.

17 *Cesar Millan*: Millan and Peltier, *Cesar's Way*.

19 *fighting breeds were crossbred*: Villavicencio, "A History of Dogfighting."

19 *"villainous-looking set"*: McCabe, *The Secrets of the Great City*, 389.

19 *"Let dogs delight to bark and bite"*: Watts, "Let Dogs Delight," 114.

19 *mercenary pet pilferers*: Information about the Fancy compiled from Woolf, *Flush*, 51, 52, 62, 109; and Mayhew, *The London Underworld in the Victorian Period*, 196.

CHAPTER 3: CAESAR III

21 *the short story "Coming, Aphrodite!"*: "Coming, Aphrodite!" originally appeared under the title "Coming, Eden Bower!" in *Smart Set* 92 (August 1920), 3–25. In addition to the title change, the magazine version was shorter and omitted the story's sexual elements. The many textual variants are listed in the appendix of *Uncle Valentine and Other Stories* edited by Bernice Slote, 177–81.

21 *"ugly but sensitive face"*: Cather, *Coming, Aphrodite! and Other Stories*, 11.

21 *"he had been bred"*: Ibid., 5.

22 *"nobody but his dog"*: Ibid., 10.

22 *"I wish you wouldn't"*: Ibid., 12.

22 *"it had never occurred"*: Ibid., 13.

22 *"But he's half the fun"*: Ibid., 25.

22 *"lying on his pallet"*: Ibid.

22   *"she had often told"*: Ibid., 33.

23   *"Caesar, lying on his bed"*: Ibid., 36.

23   *everyone who's written about this story:* See, for example, Cynthia Griffin Wolff's introduction to Cather, *Coming, Aphrodite! and Other Stories*, xxvi; and Petry, "Caesar and the Artist," 307.

24   *"never did he feel so much"*: Cather, *Coming, Aphrodite! and Other Stories*, 10.

24   *"a kind of Heaven"*: Ibid.

24   *"lost in watching"*: Ibid., 11.

24   *"accompanied by a maid"*: Wharton, *The House of Mirth*, 36.

25   *25 percent of female dog owners:* See Chomel and Sun, "Zoonoses in the Bedroom," 167–72.

25   *"How I loved that first 'Foxy' of mine"*: Adams, *Shaggy Muses*, 145.

25   *"love and understand"*: Ibid., 186.

25   *"made me into a conscious sentient person"*: Ibid., 145.

25   *strict no-pet policy:* Information about dog-friendly inner-city rail networks comes from Cohen, "Pet and Train Travel Regulations in the U.S."; European regulations compiled from www.raileurope.co.uk/Default.aspx?tabid=1662.

CHAPTER 4: ĐOUCHKA

29   *troublesome and neurotic German shepherd: Behind the Bathtub* was published first in French as *Derrière la baignoire* (Paris: Gallimard, 1962), then in English as *Behind the Bathtub: The Story of a French Dog*, translated by Peter Green (Boston: Little, Brown, 1963), and, two years later, as *Douchka: The Story of a Dog* (London: Country Book Club, 1965).

30   *"fighting off Douchka's would-be admirers"*: Audry, *Behind the Bathtub*, 170.

30   *"I could neither cure"*: Ibid., 175.

30   *"what no man had"*: Ibid., 184.

31   *"loving her gave me"*: Ibid., 43.

31   *"hope constantly springing"*: Ackerley, *We Think the World of You*, 60.

32   *"she would gaze longingly"*: Ibid.

32 *"I—I can't bear"*: Ibid., 132.

32 *"It always affected me"*: Ibid., 107.

32 *"That she was awaiting"*: Ibid., 155.

32 *"a pang goes through my heart"*: Mann, *Bashan and I*, 64.

32 *"dog love is local love"*: Garber, *Dog Love*, 14.

33 *"dogs occupy the niche"*: Knapp, *Pack of Two*, 210.

34 *"Dogs are obviously attached"*: Bradshaw, *In Defense of Dogs*, 145.

34 *"Of course it does!"*: Ibid.

34 *"I loved her"*: Ackerley, *We Think the World of You*, 107.

35 *The Prix Médicis is traditionally awarded:* Information regarding the 1962 Prix Médicis comes from Schlocker, who writes, "The disagreements among the various judges were severe this year and twice had to be determined by the deciding vote of the chairman." See Schlocker, "Literary Harvest in France," 144.

35 *"The book is easy reading"*: William, review of *Behind the Bathtub*, 297.

35 *"an angry, tormented book"*: Lewis, "Animal Farming," 29.

35 *"run-of-the-mill animal stuff"*: Nye, "In the Night Forest," 14.

35 *"presumably writes"*: Wyndham, review of *Douchka*, 708.

35 *"based on a childhood episode"*: Grellier, "Behind a Bath," 25. However, not all the reviews were bad; see, for example, Burger, "The Impact of Douchka"; and Jacobson, "She Would Have Preferred to Gnaw on a Plump Child."

36 *Past columns (all by women):* Lodge, "I Love My Dog More Than I Love My Husband"; Gibson, "I Put My Dog's Happiness First"; Griffis, "My Dog Has Outlasted All My Romantic Relationships."

36 *"it is held"*: Kuzniar, *Melancholia's Dog*, 1.

37 *"Beneath the story of"*: William, review of *Behind the Bathtub*, 297.

CHAPTER 5: EOS

39 *a beloved female greyhound belonging to Prince Albert:* Information compiled from Nichols, *The Beloved Prince*; Orpen, "Royal Favorites"; and Misztal, "Queen Victoria and Her Dogs."

39 *"She was my companion"*: Orpen, "Royal Favorites," 208.

40 *"much pleased"*: Bowater, "Queen Victoria's Silver Gift."

41 *"a darling little fellow"* and *"he had such dear"*: Marsden, *Victoria & Albert*, 270.

41 *This grotesque gewgaw:* Details of the Melbourne Centerpiece from National Gallery of Victoria collection, www.ngv.vic.gov.au/col/work/15654.

41 *"very friendly"*: Marsden, *Victoria & Albert*, 104.

41 *"Favorites often get shot"*: Misztal, "Queen Victoria and Her Dogs," 389.

42 *"dear Eos"*: Ibid.

42 *"is going on well"*: Ibid.

42 *"Eos is quite convalescent"*: Ibid.

42 *"attack"*: Ibid.

42 *"I am sure"*: Orpen, "Royal Favorites," 206.

42 *"in despair"*: *The Letters of Queen Victoria*, 474.

42 *"Poor dear Albert"*: Roberts, *Royal Artists*, 116.

43 *"Do not allow"*: Pronek, *How to Raise and Train a French Bulldog*, 14.

43 *But dogs are individuals:* Tillman, the skateboarding bulldog: www.youtube.com/watch?v=CQzUsTFqtWo; Bailey, the swimming bulldog: www.youtube.com/watch?v=rLczOGIBjHE; Sarge, the diving bulldog: www.youtube.com/watch?v=Ug2RTKH1tzU; Rosie, the motorcycling bulldog: www.youtube.com/watch?v=qwJ9_Q_vBus&ntz=1; Winston the rocking bulldog: www.youtube.com/watch?v=H6WMV_HvWKc.

44 *"in a hurry"*: Leslie, *Autobiographical Reflections*, 307.

45 *"took a severe hand"*: Stewart, *Albert*, 212.

45 *"He sees what my intentions are"*: Mann, *Bashan and I*, 61.

CHAPTER 6: FLUSH

47 *"He & I"*: *The Brownings' Correspondence*, 9:157.

47 *"given up the sunshine"*: Woolf, *Flush*, 70.

48 *"This you'll call sentimental"*: Letter to Ethel Smyth, 2 June 1935. Collected in *Letters of Virginia Woolf*, 3025.

48 *"silly"*: 29 April 1933. Collected in *Diaries of Virginia Woolf*, 153.

49 *"came up stairs"*: *Letters of Robert Browning and Elizabeth Barrett Browning*, 352.

49 *"spoken to me"*: Ibid., 361.

49 *Incidentally, Flush has no reason to complain:* Paraphrased from *Elizabeth Barrett to Miss Mitford*, 68.

49 *"I was accused":* The Brownings' Correspondence, 7:330–31.

49 *"After all it was excusable":* Ibid., 331.

50 *While he may not have ridiculed:* Information compiled from Grant, "Virginia Woolf and the Beginnings of Bloomsbury," 99.

50 *"They steal fancy dogs":* Mayhew, *The London Underworld in the Victorian Period*, 196.

50 *"the poor little creature":* J. W. Carlyle to H. Welsh, 5 June 1851, CLO 26:83–84.

50 *"for if they find I am ready":* Ibid.

51 *"and that is so sad":* Ibid.

51 *against doctor's orders:* The Brownings' Correspondence, 8:45.

51 *"we are not in the least surprised":* Freud, "The Interpretation of Dreams," 56.

53 *"with that great dog":* Barrie, *Peter Pan and Other Plays*, 78.

53 *"a look":* Ibid.

53 *"friendly and cynical mongrel":* Galsworthy, *The Forsyte Saga*, 85.

53 *"trying to be a Pomeranian":* Ibid., 301.

53 *"it is by muteness":* Galsworthy, *Some Slings and Arrows*, 42.

53 *"Family men prefer poodles":* Audry, *Behind the Bathtub*, 47.

53 *"The servant maintained":* Ibid., 63.

54 *"twa couple":* Scott, *Guy Mannering*, 119.

54 *"auld Pepper":* Ibid.

54 *"O, that's a fancy":* Ibid., 120.

54 *"not a lady's dog, you know":* Dickens, *Dombey and Son*, 179.

54 *"a blundering, ill-favored":* Ibid.

CHAPTER 7: GIALLO

57 *Landor was said to be the model:* Linton, "Reminiscences of Walter Savage Landor," 114.

58 *"all the playful":* Ibid.

58 *"Not for a million":* Forster, *Walter Savage Landor*, 427.

58 *"the old man":* Field, "The Last Days of Walter Savage Landor," 388.

58 *"The Pomeranian"*: Ouida, "Dogs and Their Affections," 315.

58 *In his lodgings:* See Field, "The Last Days of Walter Savage Landor," 544.

58 *"an approving wag"*: Ibid., 547.

59 *"He is foolish"*: Ibid., 692.

59 *"Giallo!"*: Landor, *Letters and Other Unpublished Writings*, 221.

59 *"Poor dog!"*: Ibid.

59 *"Oh Madam"*: J. W. Carlyle to E. Twisleton, 3 December 1855, CLO 30:127–129.

59 *Proust wrote regular letters:* Proust's letter cited in Grenier, *The Difficulty of Being a Dog*, 109.

60 *Clark Griswold:* See "Ultimate Dog Tease" at www.youtube.com/watch?v=nGeKSiCQkPw.

60 *Text from Dog:* http://textfromdog.tumblr.com.

60 *"a sign of affection"*: Harris, "Internet Goes to the Dogs with Blawgers." See also Tannen, "Talking the Dog."

61 *"When someone offers what sounds"*: Susan Cohen quoted in Knapp, *Pack of Two*, 109.

61 *Arnold Arluke and Clinton R. Sanders:* Quotations taken from Arluke and Sanders, *Regarding Animals*, 68–69.

CHAPTER 8: HACHIKŌ

65 *Hidesaburō Ueno, a professor of agricultural science:* Information compiled from Bondeson, *Amazing Dogs*; and Turner and Nascimbene, *Hachikō*.

66 *simply the most recent variant:* For information on the Faithful Hound motif, see Ashliman, Folklore and Mythology Electronic Texts.

67 *Like Greyfriars Bobby:* See Bondeson, *Greyfriars Bobby*.

68 *"Why did dogs"*: du Maurier, *Rebecca*, 322.

68 *"I sit outside"*: Knapp, *Pack of Two*, 41.

69 *"breeds with"*: Herzog, *Some We Love*, 112.

69 *René Descartes's claim:* See *Oeuvres des Descartes*, 243.

71 *"It struck me"*: Audry, *Behind the Bathtub*, 22.

71 *"a person's core sense"*: Knapp, *Pack of Two*, 74.

71 *"When the dog fails"*: Ibid., 66.

CHAPTER 9: ISSA

73   *"Publius' darling puppy"*: Martial, *Epigrams*, 109.

74   *"admirable"*: Ouida, "Dogs and Their Affections," 314.

74   *"gay vague"*: Colman, "Gay or Straight?" 6.

75   *"Checkers speech"*: See transcript, www.pbs.org/wgbh/americanexperience/features/primary-resources/nixon-checkers/.

75   *"Fala speech"*: See transcript, www.wyzant.com/help/history/hpol/fdr/fala.

76   *When Bush introduced Barney*: Taken from Wikipedia: http://en.wikipedia.org/wiki/Barney_%28dog%29.

79   *"too great a human"*: Audry, *Behind the Bathtub*, 231.

CHAPTER 10: JIP

81   *"showed his"*: Dickens, *David Copperfield*, 189.

81   *"Jip must have"*: Ibid., 381.

82   *"He is, as it were"*: Ibid., 283.

82   *"little petitioner"*: Eliot, *Middlemarch*, 43.

82   *"likes these small pets"*: Ibid., 44.

82   *"It is painful"*: Ibid., 43.

82   *"Ladies usually are fond"*: Ibid., 43–44.

83   *"very young ladies"*: "Pets, and What They Cost," 410.

83   *"diplomatic agents"*: Ibid., 411.

83   *"Ladies of mature age"*: Ibid.

84   *"If it had grown up"*: Carroll, *Alice's Adventures in Wonderland*, 74.

85   *"talisman"*: Bonaparte, *Topsy*, 164.

85   *"curious incident"*: Doyle, *The Complete Sherlock Holmes*, 347.

85   *"an ugly"*: Ibid., 117.

85   *"I would rather"*: Ibid., 115.

86   *Dog Betrays Woman's Infidelity*: See Ashliman, Folklore and Mythology Electronic Texts.

86   *"well-trained dog"*: *Châtelaine of Vergi*, 50–51.

86   *"The Wonder Dog"*: La Fontaine, "The Little Dog," 220–22.

87   *"I tore the bracelet"*: Douglass, *Lady Caroline Lamb*, 92.

87   *"still warm"*: Wharton, "Kerfol," 337.

87   *"her distress"*: Ibid.

87   *"dared not"*: Ibid., 338.

88   Each in His Own Way: Pirandello, *Each in His Own Way*, 13–20.

88   *"a small"*: Fitzgerald, *The Great Gatsby*, 158.

88   *"The Bitch"*: Colette, *Collected Stories*, 417–20.

CHAPTER 11: KASHTANKA

89   *"a reddish mongrel"*: Chekhov, *The Lady with the Little Dog*, trans. Wilks, 173.

89   *"Once he had even"*: Ibid., 176.

89   *"If she had been"*: Ibid., 178.

90   *"the delicious dinners"*: Ibid., 204.

90   *"a quietly sentimental tale"*: Chekhov, *Kashtanka*, trans. Meyer.

90   *"Children will respond"*: Fleming, "Review of *Kashtanka*."

90   *"particularly agonizing"*: Chekhov, "Kashtanka," trans. Garnett, 181.

90   *"Fedyushka would tie"*: Ibid.

90   *"the more lurid"*: Ibid.

91   *"she learned very eagerly"*: Ibid., 191.

91   *"the capacity for satisfaction"*: Haraway, *The Companion Species Manifesto*, 52.

93   *"Animals are more unrestrained"*: Mann, *Bashan and I*, 201.

93   *"Man himself"*: Darwin, "The Expression of the Emotions in Man and Animals," 11.

94   *"little bull-dog"*: Maeterlinck, *Our Friend the Dog*, 3.

94   *"succeeds in piercing"*: Ibid., 40.

94   *"how to preserve"*: Kuzniar, *Melancholia's Dog*, 2.

94   *"If a lion"*: Wittgenstein, *Philosophical Investigations*, 223.

94   *"To call my present idea"*: James, *Essays in Radical Empiricism*, 198.

95   *"A dog cannot lie"*: Wittgenstein, *Philosophical Investigations*, 250.

95   *"sitting at the foot"*: Maeterlinck, *Our Friend the Dog*, 65.

95   *"I envied the gladness"*: Ibid., 67.

95   *"not a single one"*: Woolf, *Flush*, 132.

95   *"we may be"*: James, *A Pluralistic Universe*, 309.

CHAPTER 12: LUMP

97   *Lump was a handsome dachshund:* Story of Lump and Picasso summarized from Duncan, *Picasso and Lump.*

99   *"squat, clumsy, deformed":* Ouida, "Dogs and Their Affections," 3.

100 *"Only someone":* Stock, "A Dog's Life."

100 *"I make no apologies":* Hockney, *Dog Days,* 5.

100 *"he quite expects it":* Ross, *The Book of Noble Dogs,* 235.

101 *"vouch'd by glorious renown":* The Poetical Works of Matthew Arnold, 467.

101 *"shining yellow coat":* Ibid., 490.

101 *"the darling himself":* Ross, *The Book of Noble Dogs,* 236.

101 *"hideously expensive":* Lustig, "James, Arnold, 'Culture,' and 'Modernity,'" 164.

101 *"a pedigree as long":* Ibid., 165.

101 *"snoring audibly":* Ibid.

102 *"the dog demanded":* National Portrait Gallery, www.npg.org.uk/ collections/search/portrait/mw08605/Benjamin-Britten.

103 *"he should sit":* The Works of George Byron, 175.

103 *the young Henry Kissinger:* Information about Kissinger and his spaniel from "Kissinger Returns."

105 *certain inmates are allowed:* See Canine Partners for Life, http:// k94life.org.

105 *85,000 Americans are injured:* See "Nonfatal Fall-Related Injuries Associated with Dogs and Cats," www.cdc.gov/mmwr/preview/ mmwrhtml/mm5811a1.htm.

CHAPTER 13: MATHE

107 *"most delicate breeds":* Shaw, *The Illustrated Book of the Dog,* 158.

108 *"the greyhound":* Froissart, *Chronicles,* 447.

108 *Disguised Man Recognized by Dog:* See Ashliman, Folklore and Mythology Electronic Texts.

109 *this dusky hellhound:* Information on spectral black dogs from Briggs, *An Encyclopedia of Fairies*, 301; and Trubshaw, ed., *Explore Phantom Black Dogs.*

110 *the beast was called the Gytrash:* See Brontë, *Jane Eyre*, 107.

110 *"the Devil will come":* Kieckhefer, *Magic in the Middle Ages*, 161.

110 *a dark dog appears:* "Cardinal Crescenzio," *Oeuvres complètes de Théodore Agrippa d'Aubigné*, 269.

110 *"The Cardinal and the Dog":* Complete Works of Robert Browning.

110 *Mephistopheles first appears to Faust:* Goethe, *Faust*, 61–62.

110 *"I passed a pleasant day":* Memoirs of Sir Walter Scott, 8:335.

110 *"when I rise":* Letters of Samuel Johnson, 2:314.

111 *Dog with Fire in Eyes:* See Ashliman, Folklore and Mythology Electronic Texts.

111 *blazing red eyes:* Information about red-eyed dogs compiled from Briggs, *An Encyclopedia of Fairies*, 301; and Trubshaw, ed., *Explore Phantom Black Dogs.*

111 *killed and eaten by his own hunting dogs:* See Ovid, *Metamorphoses*, Book 3, lines 206–35.

111 *eaten by their pets:* See Palmer, "Would Your Dog Eat Your Dead Body?"

112 *"That bosom":* Maturin, *Melmoth the Wanderer*, 236.

CHAPTER 14: NERO

115 *"It sleeps at the foot":* J. W. Carlyle to J. A. Carlyle, 10 December 1849, CLO 24:308–9.

115 *"a most affectionate":* J. W. Carlyle to J. Forster, 11 December 1849, CLO 24:309–10.

115 *"mostly bread and water":* T. Carlyle to J. W. Carlyle, 20 July 1857, CLO 32:190–92.

116 *"sprang from the library":* J. W. Carlyle to J. Welsh, 28 March 1850, CLO 25:54–56.

116 *"It was common knowledge":* Woolf, *Flush*, 139.

116 *"My little dog continues"*: J. W. Carlyle to J. Welsh, 4 March 1850, CLO 25:36–37.

116 *"more fuss"*: J. W. Carlyle to J. C. Aitkin, 10 March 1850, CLO 25:45–46.

116 *"I like him better"*: J. W. Carlyle to M. Russell, 31 December 1849, CLO 24:317–19.

117 *"comes down gloomy"*: J. W. Carlyle to J. Forster, 11 December 1849, CLO 24:309–10.

117 *"that vermin"*: J. W. Carlyle's journal, 27 March 1856, CLO 30:195–262.

117 *"to the unspeakable joy"*: J. W. Carlyle to M. Russell, 6 January 1852, CLO 27:4–5.

117 *"Dullish all of us"*: J. W. Carlyle's journal, 14 November 1855, CLO 30:195–262.

117 *"washed and combed"*: Ibid., 24 November 1855.

117 *"and was indulgent"*: J. W. Carlyle to W. Allingham, 23 February 1856, CLO 31:37–39.

117 *"Only think!"*: J. W. Carlyle to M. Russell, 20 April 1857, CLO 32:130–33.

118 *"flung him"*: T. Carlyle to J. W. Carlyle, 25 June 1859, CLO 35:122–23.

118 *"My gratitude to you"*: J. W. Carlyle to A. Barnes, 2 February 1860, CLO 36:55–56.

118 *"A passing carriage"*: Froude's Life of Carlyle, 593–94.

119 *"nobody asked"*: J. W. Carlyle to T. Carlyle, 3 July 1853, CLO 28:182–83.

119 *"by two omnibuses"*: J. W. Carlyle's journal, 24 April 1856, CLO 30:195–262.

119 *"large pound-cake"*: J. W. Carlyle to J. Welsh, 28 November 1843, CLO 17:187–91.

120 *Willie Morris's charming book: My Dog Skip*, Morris, 20, 42, 114.

CHAPTER 15: ORTIPO

123 *She named the little dog Ortipo:* English transliterations often render the bulldog's name as Ortino; *n* and *p* can be difficult to distinguish in Cyrillic handwriting; however, the Russian sources that directly quote archival documents refer to Ortipo. The point is discussed at some

length at the following source: http://forum.alexanderpalace.org/ index.php?topic=1307.165.

123   *"To tell the truth"*: Tatiana to Alexandra, 30 September 1914. Collected in Nicholas II and Alexandra, *A Lifelong Passion*, 404.

124   *Tatiana's mother was sympathetic*: Alexandra to Nicholas, 28 November 1915 and 17 March 1916, http://forum.alexanderpalace.org/index .php?topic=1307.0.

124   *"It is a very cute"*: Letter of Tatiana, 12 October 1914, http://forum .alexanderpalace.org/index.php?topic=1343.0;wap2.

124   *"They are very small"*: Tatiana to Nicholas, 17 November 1915. Collected in Zvereva, *Avgusteyshie Sestry Miloserdiya*, 36.

124   *"Ortipo had to be shown"*: Alexandra to Nicholas, 17 March 1916, http:// forum.alexanderpalace.org/index.php?topic=1343.0;wap2.

124   *"struggling to carry Ortipo"*: Bokhanov et al., *The Romanovs*, 310.

124   *Finally, the poor creature's body*: See King and Wilson, *The Fate of the Romanovs*, 276, 285, 312.

125   *historians are unsure*: This question is debated in some detail at the following forum: http://forum.alexanderpalace.org/index.php?topic =1307.5;wap2.

125   *soldiers impaling babies*: See "The Bayonet Baby Effect," http://antzml .wordpress.com/2011/02/25/the-bayonet-baby-effect.

125   *victorious display of a royal dog*: See "Early History of the Breed," www .rosebury.de/breedhistory.html.

126   *"all I observed there was the silliness"*: Ibid.

126   *"dangerously, irksomely and horribly"*: See "King Edward VI," http:// englishhistory.net/tudor/monarchs/edward6.html.

127   *Faithful Lapdog*: See Ashliman, Folklore and Mythology Electronic Texts.

127   *"The Queen's head fell"*: "Pups of the Past," http://marie antoinettequeenoffrance.blogspot.com/2011/08/pups-of-past-marie -antoinettes-dogs.html.

127   *audaciously honest name Looty*: See http://en.wikipedia.org/wiki/ Pekingese.

128 *Notoriously, these charming little creatures:* History of the breed summarized at the website of the French Bull Dog Club of America, http://frenchbulldogclub.org/about-frenchies/understanding-frenchies/understanding-the-breeds-history.

130 *"a real Parisian guttersnipe":* Felix Yusupov, *Lost Splendor* 5, www.alexanderpalace.org/lostsplendor/v.html.

130 *"best model":* Ibid.

130 *"unpleasant looking":* Chukovskaia, *To the Memory of Childhood,* 72.

131 *He survived the shipwreck:* See "Frenchies and the Titanic," French Bull Dog Club of America, http://frenchbulldogclub.org/about/our-clubs-history/frenchies-and-the-titanic.

131 *still nowhere near as fashionable:* Rankings listed at www.akc.org/reg/dogreg_stats.cfm.

132 *"We English":* See www.diplomate-ambassadeur.cz/en/french-bulldog.

132 *"the French Bulldog originated as":* See www.akc.org/breeds/french_bulldog/index.cfm.

CHAPTER 16: PERITAS

135 *Peritas, the favorite dog of Alexander:* From Plutarch, *Life of Alexander,* 61:3.

136 *"guardian of the herds":* Aristotle, *History of Animals,* 9:1.

136 *"Never, with them":* Virgil, *Georgics,* 3:404ff.

137 *"the true Molossian":* Wynn, *The History of the Mastiff Breed,* 34.

137 *one of the first joke dog names:* See Petronius, *Satyricon,* 142–45.

137 *a deer in the amphitheater:* Martial, *Epigrams.*

137 Cave Canem: Wynn, *The History of the Mastiff Breed,* 34.

137 *bravery of the* pugnaces britanniae*:* Grattius, "Cynegetica," 179ff.

138 *"hounds and greyhounds":* Shakespeare, *Macbeth,* 3.1.95–104.

138 *adopting a rescue mutt:* The trend for rescue dogs is chronicled by Jodi Lastman at the blog Hypenotic, http://hypenotic.com/rescue-dogs-a-trend-unleashed/.

139 *the most expensive dog:* See McGraw, "$1.5 Million Paid for World's Most Expensive Dog."

139 *"meets the standards":* See http://en.wikipedia.org/wiki/Dog_breed.

139 *the visual demands of a "perfect specimen":* See the American Kennel Club glossary at www.akc.org/about/glossary.cfm and "Dog Show Lingo" at http://nefer-temu.8m.com/info/glossary.htm.

139 *"We may sometimes":* See http://spoiledmaltese.com/forum/59-everything-else-maltese-related/114583-genetic-memory.html.

140 *The Pekingese is a good case in point:* See Cheang, "Women, Pets and Imperialism."

CHAPTER 17: QUININE

143 *Anton Chekhov was promised two puppies:* Description from Rayfield, *Anton Chekhov,* 293.

144 *"The dachshunds have been":* Ibid.

144 *Masha named the male dog:* Ibid.

144 *"Brom is nimble":* Ibid.

144 *"every evening":* Ibid.

144 *"Gurov was on the point":* Chekhov, *The Lady with the Little Dog and Other Stories,* 234.

145 *"In the family albums":* Nabokov, *Speak, Memory,* 30.

145 *"grizzled muzzle":* Ibid., 85.

145 *"because of his being":* Ibid., 30.

145 *"until he was chloroformed":* Ibid.

146 *"one of my few connections":* Updike, "Nabokov's Look Back," 15.

146 *"he could be still seen":* Nabokov, *Speak, Memory,* 30.

146 *"fat yellow dachshund":* Nabokov, *Laughter in the Dark,* 30–31.

146 *"dropsical dackel":* Nabokov, *Lolita,* 206.

146 *"turned up":* Nabokov, *Ada, or Ardor,* 233.

147 *A notorious accident:* Story of Newton and Diamond told in Coren, *The Pawprints of History,* 292–93.

147 *"broad and hairy paws":* Mann, *Bashan and I,* 56.

147 *"beautifully formed":* The Works of George Byron, 114.

148 *"he and his master":* Ibid.

148 *"he has already bitten":* Ibid., 159.

148 *"My bull-dog is deceased":* Ibid., 159–60.

148   *"How is . . . the Phoenix"*: Ibid., 169.

149   *"as many more"*: Ibid., 215.

149   *"Scott kept one window"*: Chambers and Chambers, "Sir Walter Scott and His Dogs," 274.

150   *"It is very awful"*: This and other details about Dickens's dogs from "Dogs of Literature," 493.

150   *For these reclusive types*: For more about Dickinson, Brontë, and their dogs, see Adams, *Shaggy Muses*.

CHAPTER 18: ROBBER

153   *"this experiment"*: Wagner, *My Life*, vol. 1, www.gutenberg.org/cache/epub/5197/pg5197.html.

154   *"so touched the hearts"*: Ibid.

154   *"moved to pity"*: Ibid.

155   *"irreconcilable dislike"*: Ibid.

155   *"crosswise"*: Ibid.

155   *"must have run away"*: Ibid.

155   *"I have always considered"*: Ibid.

156   *"seemed only to revive"*: Ibid.

156   *"The fact that he"*: Ibid.

157   *"like a little dog"*: This and other details of Wagner's dreams from Köhler, *Richard Wagner*, 551.

158   *"Little Dog Waltz"*: See http://worldofopera.org/component/k2/item/450-waltz-in-d-flat-major-op-64-no-1.

158   *known as the Leskovites*: See http://bibliolore.org/2012/03/05/busoni-and-the-leskovites.

158   *"was more agreeable company"*: See www.wisdomportal.com/RachmaninoffNotes.html.

158   *"docile, and always sagacious"*: Shaw, *Illustrated Book of the Dog*, 64.

158   *"the effete miniature dog"*: Kogan, "From Russia, with Love."

158   *"I love Pooks"*: Ibid.

158   *"for a dog"*: Volta, "Give a Dog a Bone."

159   *Laurie Anderson's* Music for Dogs: A performance of the work may be seen on video at www.youtube.com/watch?v=38g4VzkIf14.

159 *Elgar's* Enigma Variations: For Edward Elgar's notes on the *Enigma Variations*, see http://en.wikipedia.org/wiki/Enigma_Variations.

159 *"Moods of Dan"*: See www.elgar.org/2theman.htm.

160 The Metamorphosis of Dan: See http://en.wikipedia.org/wiki/George _Robertson_Sinclair.

162 *"It is best"*: Pronek, *How to Raise and Train a French Bulldog*, 27.

162 *"quaint, cosy"*: Mann, *Bashan and I*, 28.

CHAPTER 19: SHOCK

166 *"Shock, who thought"*: Pope, "The Rape of the Lock," canto 1, line 115.

166 *"May Kiss"*: Brown, *Homeless Dogs and Melancholy Apes*, 72.

166 *"downy breast"*: Ibid., 71.

166 *"Securely on her Lap"*: Ibid., 72.

167 *Cardinal Paleotti wrote a decree*: Langdon, *Medici Women*, 155.

167 *"politely concealed"*: Mitchell, cited in Drabble, *The Pattern in the Carpet*, 230.

167 *"Dance all night"*: Pope, "The Rape of the Lock," canto 5, line 19.

167 *"dire Disaster"*: Ibid., canto 2, line 103.

167 *"stain her Honor"*: Ibid., line 107.

167 *"lose her Heart"*: Ibid., line 109.

167 *"that* Shock *must fall"*: Ibid., line 110.

167 *"Guard of Shock"*: Ibid., line 116.

167 *"When Husbands"*: Ibid., canto 3, line 158.

167 *"No more thy hand"*: The Works of Mr. John Gay, 94.

168 *"In man"*: Ibid., 95.

168 *"Now Clo's soft skin"*: Fothergill, "On the Premature Death of Cloe Snappum," 249.

168 *"satisfie the delicateness"*: Caius, *Of Englishe Dogges*, 20.

168 *"This abuse"*: Ibid.

169 *Chaucer's flirty, foolish Prioress*: Chaucer, *The Canterbury Tales*, 6.

169 *even her lapdogs turn against him*: Shakespeare, *King Lear*, 290.

169 *After Joséphine married Napoléon*: Stuart, *The Rose of Martinique*, 133, 198, 210.

170 *"sitting and calling"*: Austen, *Mansfield Park*, 50.

170 *"thinking more"*: Ibid., 12.

170 *"she cared for her dog"*: Woolf, *Mrs. Dalloway*, 10.

170 *a female Great Dane named Bounce*: Details about Pope and Bounce compiled from Johnson, "The Life of Pope."

171 *Shorty Rossi and his team of little people*: Pit Boss is shown on Animal Planet; see http://animal.discovery.com/tv-shows/pit-boss/bios/shorty-rossi.htm.

172 *"Don't let that dog"*: Ackerley, *My Dog Tulip*, 5.

CHAPTER 20: TULIP

176 *"rather wooden terrier"*: Ackerley, *My Dog Tulip*, 92.

177 *"I realized clearly"*: Ibid.

177 *"Are not its ghosts"*: Ibid., 37.

177 *"What's the bleeding street"*: Ibid., 33.

177 *those regarding defecation are more recent*: See Brandow, *New York's Poop Scoop Law*.

178 *Dogipots*: www.dogipot.com.

178 *DoodyDanglers*: http://doodydangler.com.

178 *Poop-Freeze*: www.poopfreeze.com.

179 *difficulty pronouncing the words*: See www.allshihtzu.com/How_To_Pronounce_Shih_Tzu.html.

179 *"you have to start"*: Audry, *Behind the Bathtub*, 45.

183 *"sense of disgust"*: Derges et al., "Complaints About Dog Faeces," 419.

CHAPTER 21: ULISSES

185 *"Ulisses is mixed-race"*: Cited in Moser, *Why This World*, 332.

185 *"I and my dog"*: Lispector, *A Breath of Life*, 51.

185 *"I'm a little impolite"*: Cited in Moser, *Why This World*, 331.

185 *"smokes cigarettes"*: Ibid., 332.

186 *"calmly let him do"*: Ibid.

186 *"One look at him"*: Cited ibid., 159.

186   *"he looked so happy"*: Ibid.

186   *"a rather grand name"*: Lispector, *Selected Crônicas*, 170.

186   *"Dilermando liked"*: Cited in Moser, *Why This World*, 159.

186   *the Swiss didn't allow dogs*: Information about Swiss hotels given in Lispector, *Selected Crônicas*, 174.

186   *"I don't like"*: Moser, *Why This World*, 165.

187   *"There are so many ways"*: Cited in ibid., 166.

188   *"there was not a wild beast"*: Homer, *The Odyssey*, book 17.

188   *"full of fleas"*: Ibid.

188   *"he dropped his ears"*: Ibid.

188   *According to French Kennel Club rules*: See www.braquedubourbonnais .info/en/dog-name.htm; also Grenier, *The Difficulty of Being a Dog*, 60.

188   *after Alfred Jarry's play* Ubu Roi*:* See Grenier, *The Difficulty of Being a Dog*, 60.

189   *"old dog"*: Cited in Adams, *Shaggy Muses*, 44.

189   *"an old mangy creature"*: Ibid.

190   *"mange-ridden"*: Audry, *Behind the Bathtub*, 47.

190   *"slobbering, overweight"*: Ibid., 48.

190   *"slobbering mongrels"*: Ibid.

191   *"When you love a dog"*: Grenier, *The Difficulty of Being a Dog*, 29.

191   *"A dog's life"*: Bonaparte, *Topsy*, 85.

191   *"She was 95"*: Grenier, *The Difficulty of Being a Dog*, 29.

CHAPTER 22: VENOM

193   *"A tiger perhaps"*: Berkeley, *My Life and Recollections*, 182.

194   *"She has got an antelope"*: *The Journal of Mrs. Arbuthnot*, 36.

194   *"a peculiar"*: Vane, preface to *Memoirs and Correspondence of Viscount Castlereagh*, 82.

194   *"Contradicting their looks"*: Rush, *Memoranda of a Residence at the Court of London*, 51.

194   *"Two pet dogs"*: Ibid., 185.

194   *"sharing the small space"*: Brownlow, *Slight Reminiscences of a Septuagenarian*, 67.

194   *"the sinews of the first"*: "Lord Castlereagh," 548.

195   *"lest my ill will"*: Cited in Bew, *Castlereagh*, 440.

195   *"a favorite dog"*: "Lord Castlereagh," 548.

195   *"had hitherto overwhelmed"*: *Memoirs of the Comtesse de Boigne*, 187.

195   *"It was not until"*: Ibid.

195   *a satirical pamphlet*: Hone, *Official Account of the Noble Lord's Bite!*.

195   *"driving in an open carriage"*: Meyrick, *House Dogs and Sporting Dogs*, 65.

196   *"The bull-dog is scarcely"*: Youatt, *The Dog*, 98.

196   *"speedily become profligate"*: Youatt, *The Obligation and Extent of Humanity to Brutes*, 169.

196   *"Who will believe"*: Darwin, *On the Origin of Species*, 36–37.

196   *"Everyone I know"*: Haraway, *The Companion Species Manifesto*, 26.

196   *The bulldog story*: Bulldog history compiled from Caius, *Of Englishe Dogges*; and Shaw, *The Illustrated Book of the Dog*.

197   *"Of all the dogs"*: *The Poems of the Late Christopher Smart*, 6.

198   *"Of all dogs"*: "A Chapter on Dogs," 41.

CHAPTER 23: WESSEX

201   *"Two things"*: Orel, *The Final Years of Thomas Hardy*, 87.

201   *"My husband"*: *Letters of Emma and Florence Hardy*, 247.

202   *"Florence's unspeakable dog"*: Tomalin, *Thomas Hardy*, 317.

202   *"the most despotic dog"*: Cited in Millgate, *Thomas Hardy*, 489.

202   *On Christmas Day, Wessex*: Ibid.

203   *"Wessie sends his love"*: *Letters of Emma and Florence Hardy*, 170.

203   *getting up early*: Tomalin, *Thomas Hardy*, 340.

203   *Henry Watkins, whom he welcomed*: Anecdote about Watkins told in Flower, "Walks and Talks with Thomas Hardy," 231–32.

204   *"We miss him"*: *Letters of Emma and Florence Hardy*, 170.

204   *"Of course he was"*: Ibid., 247.

204   *"But I mustn't write"*: Ibid.

205   *"Excessive dog love"*: Gopnik, "Dog Story," 47.

205   *"the pathos-laden presence"*: Baudrillard, "The System of Collecting," 10.

205   *"Don't all loves"*: Garber, *Dog Love*, 93.

205 *Edmund Leach has a different explanation:* See Serpell, *In the Company of Animals*, 67.

206 *"who is evidently":* Cited in Tomalin, *Thomas Hardy*, 362.

206 *"beset with desire":* Ibid., 363. See also Richardson, "The Many-Sided Thomas Hardy."

CHAPTER 24: XOLOTL

209 *the doglike deity Xolotl:* Summary of the Xolo's history from the American Kennel Club, www.akc.org/breeds/xoloitzcuintli/history.cfm.

210 *the Xolo's importance:* Xolo breed trends summarized from Judie Smith, "Mirasol Xoloitzcuintli," http://mirasolxolos.webs.com/xolohistory.htm.

211 *he was also a rebel, like his mistress:* Señor Xolotl's antics described in Tibol, *Frida Kahlo*, 158.

212 *Sightings of them:* White, *Myths of the Dog-Man*, 1–20.

213 *"kind of like":* *Fortean Times*, April 2012, 72.

213 *"In Yorkshire folklore":* Ibid.

213 *"the actual":* Hole, *Witchcraft in England*, 40.

213 *familiars usually take the form:* Details ibid., 35–45.

214 *"fiendish servants":* Kunze, *Highroad to the Stake*, 386.

215 *"lesser demon":* Russell, *Witchcraft in the Middle Ages*, 191.

215 *"were often given":* Wilby, *Cunning Folk and Familiar Spirits*, 60–63.

216 *people do prefer dogs that look similar:* Coren, "Do Dogs Look Like Their Owners?"

216 *"French bullgod":* http://tatianaromanova.piczo.com/ortino?cr=5&linkvar=000044.

CHAPTER 25: YOFI

217 *"One can love":* Beck and Katcher, *Between Pets and People*, 127.

218 *"with a high level":* Grinker, *Fifty Years in Psychiatry*, 9.

218 *"high-risk dogs":* In a study in the *Journal of the American Veterinary Medical Association*, out of 238 fatalities related to dog bites from 1979 to 1998, chows were responsible for eight. See Sacks, Sinclair, and Gilchrist, "Breeds of Dogs Involved in Fatal Human Attacks."

218   *"She is a charming creature":* Cited in Green, "Freud's Dream Companions," 67.

219   *"a little lion-like creature":* H.D., *Tribute to Freud*, 98.

219   *"passionate affection":* Bonaparte, *Topsy*, 37.

219   *"proudly erect":* Ibid., 80.

220   *Freud and Yofi would eat together:* Information about Freud's and Yofi's lunch from Green, "Freud's Dream Companions," 67.

220   *"How does a little animal":* H.D., *Tribute to Freud*, 179.

220   *"I wish you could":* Cited in Edmundson, *The Death of Sigmund Freud*, 91.

220   *"One cannot easily get over":* Ibid., 92.

220   *"animal asylum":* The Diary of Sigmund Freud, 252.

221   *"nothing could have kept":* Ibid.

221   *"I have never seen":* Ibid.

221   *his jaw gave off a putrid smell:* Edmundson, *The Death of Sigmund Freud*, 213–14.

222   *"I am impressed":* Woodward, "Lucian Freud's Whippet."

222   *Dorothy Burlingham had a gray Bedlington:* H.D., *Tribute to Freud*, 174.

222   *her cocker spaniel Butschi:* See Hitchcock, *Karen Horney*, 93.

222   *Lacan said that his boxer:* See "The Seminar of Jacques Lacan," 26.

223   *Harry Stack Sullivan:* See Clinebell, *Contemporary Growth Therapies*, 82.

223   *recueillement:* See von Armin, *All the Dogs of My Life*, 84.

CHAPTER 26: ZÉMIRE

225   *Catherine the Great, whose favorite greyhound:* Information compiled from Ross, *The Book of Noble Dogs*, 88–89.

226   *the talent of her dog:* Described in ibid., 137–39.

226   *Chaser, a border collie trained:* Nosowitz, "I Met the World's Smartest Dog."

228   *Greek philosopher Diogenes:* Laërtius, *The Lives and Opinions of Eminent Philosophers*, 231–32.

228   *"the most philosophical":* Plato, *The Republic*, 1.2.

228   *"Did you ever":* Rabelais, *Gargantua and Pantagruel*, prologue.

229   *"signs on paper":* Bonaparte, *Topsy*, 80.

229   *"whose happiness is confined":* Ibid.

Bibliography

Abbott, Frank Frost. *Society & Politics in Ancient Rome*: *Essays & Sketches*. New York: Charles Scribner's Sons, 1909.

Ackerley, J. R. *My Dog Tulip*. New York: New York Review Book Classics, 1999.
———. *We Think the World of You*. New York: New York Review Book Classics, 2012.

Adams, Maureen. *Shaggy Muses*: *The Dogs Who Inspired Virginia Woolf, Emily Dickinson, Elizabeth Barrett Browning, Edith Wharton, and Emily Brontë*. New York: Random House, 2009.

Alexander, Carolyn. *Bull Terriers*. Hauppauge, New York: Barron's Educational Series, 2006.

Arbuthnot, Harriet. *The Journal of Mrs. Arbuthnot, 1820–1832*, vol. 1. London: Macmillan, 1950.

Aristotle. *History of Animals*. Translated by D'Arcy Wentworth Thompson. London: John Bell, 1907, http://classics.mit.edu/Aristotle/history_anim.9.ix.html.

Arluke, Arnold, and Clinton R. Sanders. *Regarding Animals*. Philadelphia: Temple University Press, 1996.

Arnold, Matthew. *The Poetical Works of Matthew Arnold*. Whitefish, MT: Kessinger Publishing Company, 2005.

Ashliman, D. L. Folklore and Mythology Electronic Texts, University of Pittsburgh, www.pitt.edu/~dash/folktexts.html.

Aubigné, Théodore-Agrippa d'. *Oeuvres complètes de Théodore Agrippa d'Aubigné*. Edited by A. Legouëz. Paris: A. Lemerre, 1891.

Audry, Colette. *Behind the Bathtub*: *The Story of a French Dog*. Translated by Peter Green. Boston: Little, Brown, 1963 (English edition published as *Douchka*: *The Story of a Dog*, Souvenir Press, 1963).

Austen, Jane. *Mansfield Park*. London: W. W. Norton, 1999.

Barrie, J. M. *Peter Pan and Other Plays*. Oxford University Press, 1999.

Baudrillard, Jean. "The System of Collecting." In *The Cultures of Collecting*, edited by Roger Cardinal and John Elsner, 7–24. London: Reaktion Books, 1994.

Beck, Alan, and Aaron Katcher. *Between Pets and People*: *The Importance of Animal Companionship*. West Lafayette, IN: Purdue University Press, 1996.

Benfer, Amy. "Gertrude and Alice." *Salon*, 18 November 1999, www.salon.com/1999/11/18/alice/.

Berkeley, Grantley Fitzhardinge. *My Life and Recollections*, vol. 3. London: Hurst and Blackett, 1866.

Bew, John. *Castlereagh*: *A Life*. Oxford University Press, 2012.

Boigne, Louise-Eléonore-Charlotte-Adélaide d'Osmond, Comtesse de. *Memoirs of the Comtesse de Boigne*, vol. 2. Edited by Charles Nicoullaud. Paris: C. Scribner's Sons, 1908.

Bokhanov, Alexander, Dr. Manfred Knodt, Vladimir Oustimenko, Zinaida Peregudova, and Lyu Tyutyunnik. *The Romanovs*: *Love, Power, and Tragedy*. Translated by Lyudmila Xenofontova. London: Leppi Publications, 1993.

Bonaparte, Marie. *Topsy*: *The Story of a Golden-Haired Chow*. Edison, NJ: Transaction Books, 1994.

Bondeson, Jan. *Amazing Dogs*: *A Cabinet of Canine Curiosities*. Ithaca, NY: Cornell University Press, 2011.

———. *Greyfriars Bobby*: *The Most Faithful Dog in the World*. Stroud, UK: Amberley, 2011.

Bowater, Donna. "Queen Victoria's Silver Gift to Prince Albert on Sale for £200,000." *Telegraph*, 13 June 2012, www.telegraph.co.uk/news/uknews/theroyalfamily/9327775/Queen-Victorias-silver-gift-to-Prince-Albert-on-sale-for-200000.html.

Bradshaw, John. *Dog Sense*: *How the New Science of Dog Behavior Can Make You a Better Friend to Your Pet*. New York: Basic Books, 2011.

———. *In Defense of Dogs: Why Dogs Need Our Understanding*. London: Allen Lane, 2011.

Brandes, Stanley. "The Meaning of American Pet Cemetery Gravestones." *Ethnology* 48, no. 2 (Spring 2009), 99–118.

Brandow, Michael. *New York's Poop Scoop Law*: *Dogs, the Dirt, and Due Process*. West Lafayette, IN: Purdue University Press, 2008.

Briggs, Katharine Mary. *An Encyclopedia of Fairies.* London: Pantheon Books, 1976.

Brontë, Charlotte. *Jane Eyre.* London: J.M. Dent & Sons Ltd., 1922.

Brown, Laura. *Homeless Dogs and Melancholy Apes: Humans and Other Animals in the Modern Literary Imagination.* Ithaca, NY: Cornell University Press, 2010.

Browning, Elizabeth Barrett. *Elizabeth Barrett to Miss Mitford: The Unpublished Letters of Elizabeth Barrett to Mary Russell Mitford.* Edited by Betty Miller. London: John Murray, 1954.

————. *The Letters of Elizabeth Barrett.* Edited by Frederic G. Kenyon. London: Macmillan, 1899.

Browning, Robert. *Complete Works of Robert Browning*, vol. 12. London: T. Y. Crowell, 1898.

Browning, Robert, and Elizabeth Barrett Browning. *The Brownings' Correspondence*, vols. 5–8. Edited by Philip Kelley and Ronald Hudson. Waco, TX: Wedgestone Press, 1984–87.

————. *The Brownings' Correspondence*, vols. 9–14. Edited by Philip Kelley and Scott Lewis. Waco, TX: Wedgestone Press, 1984–87.

————. *The Letters of Robert Browning and Elizabeth Barrett Browning, 1845–1846: With Portraits and Facsimiles*, vol. 2. Edited by Robert Browning. New York: Harper & Bros., 1899.

Brownlow, Emma Sophia Edgcumbe Cust, Countess. *Slight Reminiscences of a Septuagenarian from 1802 to 1815.* London: John Murray, 1867.

Burger, Otis K. "The Impact of Douchka." *New York Times*, 24 November 1963, 207.

Byron, George Gordon, Baron. *The Works of George Byron: With His Letters and Journals, and His Life*, vol. 1. Edited by Thomas Moore. London: John Murray, 1835.

Caius, John. *Of Englishe Dogges.* Translated by Abraham Fleming. Project Gutenberg (EBook #27050), www.gutenberg.org/files/27050/27050-h/27050-h.htm.

Camus, Albert. *The Stranger.* Translated by Matthew Ward. New York: Random House, 2012.

Carlyle, Thomas, and Jane Welsh Carlyle. *The Carlyle Letters Online* (CLO). Edited by Brent E. Kinsner. Duke University Press, 14 September 2007.

Carroll, Lewis. *Alice's Adventures in Wonderland.* Wellesley, MA: Branden Books, 1947.

Cartwright, David E. *Historical Dictionary of Schopenhauer's Philosophy.* Lanham, MD: Scarecrow Press, 2005.

Cather, Willa. *Coming, Aphrodite! and Other Stories.* Edited by Margaret Anne O'Connor. New York: Penguin, 1999.

Chambers, William, and Robert Chambers. "Sir Walter Scott and His Dogs." *Chambers's Journal,* 5 May 1877, 273–76.

"A Chapter on Dogs." *The Illustrated London Reading Book.* London, 1851. Project Gutenberg (EBook #11921), www.gutenberg.org/files/11921/11921 -h/11921-h.htm.

Châtelaine of Vergi, The: A 13th Century French Romance. Translated by Alice Kemp-Welch. London: D. Nutt, 1903.

Chaucer, Geoffrey. *The Canterbury Tales.* Translated by David Wright. Oxford University Press, 2011.

Cheang, Sarah. "Women, Pets and Imperialism: The British Pekingese Dog and Nostalgia for Old China." *Journal of British Studies* 45, no. 2 (April 2006), 359–87.

Chekhov, Anton. "Kashtanka." *The Tales of Chekhov.* Vol. 12, *The Cook's Wedding and Other Stories.* Translated by Constance Garnett. New York: Ecco Editions, 1986.

———. *Kashtanka.* Translated by Ronald Meyer. Illustrated by Gennady Spirin. Boston: HMH Books for Young Readers, 1995.

———. *The Lady with the Little Dog and Other Stories, 1896–1904.* Translated by Ronald Wilks. London: Penguin, 2002.

Chomel, Bruno B., and Ben Sun. "Zoonoses in the Bedroom." *Emerging Infectious Diseases* 17, no. 2 (February 2011).

Christie, Agatha. *Dumb Witness.* New York: HarperCollins, 2005.

Chukovskaia, Lidia Korneevna. *To the Memory of Childhood.* Evanston, IL: Northwestern University Press, 1988.

Clinebell, Howard John. *Contemporary Growth Therapies: Resources for Actualizing Human Wholeness.* Nashville, TN: Abingdon Press, 1981.

Cohen, Gail. "Pet and Train Travel Regulations in the U.S." Travel Tips, *USA Today,* http://traveltips.usatoday.com/pet-train-travel-regulations -us-15601.html.

Colette [Sidonie-Gabrielle Colette]. *The Collected Stories of Colette.* Edited by Robert G. Phelps. Translated by Antonia White, Matthew Ward, and Anne-Marie Callimachi. New York: Farrar, Straus and Giroux, 1984.

Colman, David. "Gay or Straight? Hard to Tell." *New York Times,* 19 June 2005, 6.

Coren, Stanley. "Do Dogs Look Like Their Owners?" *Psychology Today,* 7 August 2013, www.psychologytoday.com/blog/canine-corner/201308/do-dogs-look-their-owners.

—————. *The Pawprints of History: Dogs in the Course of Human Events.* New York: Simon and Schuster, 2002.

Darwin, Charles. "The Expression of the Emotions in Man and Animals." In *Works of Charles Darwin.* London: D. Appleton, 1896.

—————. *On the Origin of Species.* London: P.F. Collier and Son, 1909.

Derges, Jane, Rebecca Lynch, Angela Clow, Mark Petticrew, and Alizon Draper. "Complaints About Dog Faeces as a Symbolic Representation of Incivility in London, UK: A Qualitative Study." *Critical Public Health* 22, no. 4 (December 2012), 419–25.

Descartes, René. *Oeuvres de Descartes.* Edited by C. Adam and P. Tannery. Paris: L. Cerf, 1897–1913.

Dickens, Charles. *David Copperfield.* Leipzig: Tauschnitz, 1850.

—————. *Dombey and Son.* London: Bradbury and Evans, 1848.

—————. *Little Dorrit.* New York: G.W. Carleton, 1885.

—————. *Oliver Twist.* Boston: Ticknor and Fields, 1866.

—————. *Oliver Twist, or The Parish Boy's Progress.* Illustrated by George Cruikshank. London: Richard Bentley, 1838.

"Dogs of Literature." *Temple Bar* 61 (January–April 1881), 476–500.

Douglass, Paul. *Lady Caroline Lamb: A Biography.* New York: Palgrave Macmillan, 2004.

Doyle, Sir Arthur Conan. *The Complete Sherlock Holmes.* New York: Bantam Classics, 1986.

Drabble, Margaret. *The Pattern in the Carpet: A Personal History with Jigsaws.* Boston: Houghton Mifflin Harcourt, 2010.

du Maurier, Daphne. *Rebecca.* New York: William Morrow, 2006.

Duncan, David Douglas. *Picasso and Lump: A Dachshund's Odyssey.* New York: Bullfinch Press, 2006.

Edmundson, Mark. *The Death of Sigmund Freud: The Legacy of His Last Days*. New York: Bloomsbury, 2007.

Eliot, George. *Middlemarch*. London: William Blackwood, 1871.

Field, Kate. "The Last Days of Walter Savage Landor," pts. 1, 2, and 3. *Atlantic Monthly*, April 1866, 385–95; May 1866, 540–52; June 1866, 684–705.

Fischer, Renaldo. *The Shaman Bulldog: A Love Story*. New York: Warner Books, 1996.

Fitzgerald, F. Scott. *The Great Gatsby*. New York: Scribner, 2004.

Fleming, Jennifer. "Review of *Kashtanka*, by Anton Chekhov, Translated by Ronald Meyer, Illustrated by Gennady Spirin." *School Library Journal*, 1 December 1995, 73.

Flower, Newman. "Walks and Talks with Thomas Hardy." In *Thomas Hardy Remembered*, edited by Martin Ray, 231–32. Burlington, VT: Ashgate Publishing, 2007.

Forster, John. *Walter Savage Landor: A Biography*. London: Chapman and Hall, 1869.

Fothergill, Anthony. "On the Premature Death of Cloe Snappum, A Lady's Favorite Lap-Dog." In *An Asylum for Fugitive Pieces, in Prose and Verse, Not in Any Other Collection: with Several Pieces Never Before Published. A New Ed., Including Pieces Not in the Former Edition, and Several Never Before Printed*, vol. 3, 248–49. London: Debrett, 1795.

Freud, Sigmund. *The Diary of Sigmund Freud: A Record of the Final Decade*. Edited and translated by Michael Molnar. New York: Scribner, 1992.

———. "The Interpretation of Dreams" (first part). In *The Standard Edition of the Complete Psychological Works of Sigmund Freud*, vol. 4, i–xiii. Translated by James Strachey. London: The Hogarth Press and the Institute of Psycho-analysis, 1900.

Froissart, Jean. *Chronicles*. Edited by Geoffrey Brereton. London: Penguin Classics, 1978.

Froude, James Anthony. *Froude's Life of Carlyle*. Edited by John Clubbe. Columbus: Ohio State University Press, 1979.

Galsworthy, John. *The Forsyte Saga*. Edited by Geoffrey Harvey. New York: Oxford University Press, 2008.

———. *Some Slings and Arrows*. Hong Kong: Forgotten Books, 2012.

Garber, Marjorie. *Dog Love*. New York: Simon and Schuster, 1996.

Gay, John. *The Works of Mr. John Gay: In Four Volumes. To Which Is Added an Account of the Life and Writings of the Author.* Dublin: James Potts, 1770.

Gibson, Crystal. "I Put My Dog's Happiness First." *Dogster,* 18 March 2013, www.dogster.com/lifestyle/my-dog-saved-my-marriage.

Goethe, Johann Wolfgang von. *Faust.* Translated by Charles T. Brooks. Boston: Ticknor and Fields, 1856.

Gopnik, Adam. "Dog Story." *New Yorker,* 8 August 2011, 46–52.

Grant, Duncan. "Virginia Woolf and the Beginnings of Bloomsbury." In *The Bloomsbury Group: A Collection of Memoirs and Commentary,* edited by Stanford Patrick Rosenbaum, 97–102. Toronto: University of Toronto Press, 1995.

Grattius. "Cynegetica." In *Minor Latin Poets,* vol. 1, translated by J. Wight Duff and Arnold M. Duff, 284. Cambridge, MA: Harvard University Press, 1935.

Green, Susie. "Freud's Dream Companions." *Guardian,* 22 March 2002, www.theguardian.com/theguardian/2002/mar/23/weekend7 .weekend3.

Grellier, Maryvonne. "Behind a Bath." *Guardian,* 27 October 1990, 25.

Grenier, Roger. *The Difficulty of Being a Dog.* Translated by Alice Kaplan. University of Chicago Press, 2000.

Griffis, Gigi. "My Dog Has Outlasted All My Romantic Relationships." *Dogster,* 11 March 2013, www.dogster.com/lifestyle/my-dog-has-outlasted -romantic-relationships.

Grinker, Roy Richard. *Fifty Years in Psychiatry: A Living History.* Springfield, IL: Charles C. Thomas, 1979.

Haraway, Donna. *The Companion Species Manifesto: Dogs, People, and Significant Otherness.* Chicago: Prickly Paradigm Press, 2003.

Hardy, Florence, and Emma Hardy. *Letters of Emma and Florence Hardy.* London: Oxford University Press, 1996.

Harris, Misty. "Internet Goes to the Dogs with Blawgers." Canada .com, date unknown. Article no longer available but cited at http:// terriorists.blogspot.com/2004/07/dogs-who-blog.html.

H.D. [Hilda Doolittle]. *Tribute to Freud.* New York: New Directions, 1974.

Herzog, Hal. *Some We Love, Some We Hate, Some We Eat.* New York: Harper, 2010.

Hitchcock, Susan Tyler. *Karen Horney.* New York: InfoBase, 2009.

Hockney, David. *Dog Days.* London: Thames and Hudson, 2006.

Hole, Christina. *Witchcraft in England.* London: Rowan and Littlefield, 1977.

Homer. *The Odyssey.* Translated by Samuel Butler. Project Gutenberg (EBook #1727), www.gutenberg.org/files/1727/1727-h/1727-h.htm.

Hone, William. *Official Account of the Noble Lord's Bite! and His Dangerous Condition, with Who Went to See Him, and What Was Said, Sung, and Done, on the Melancholy Occasion. Published for the Instruction and Edification of All Ranks and Conditions of Men by the Author of Buonaparte-Phobia; or, Cursing Made Easy.* London: W. Hone, 1817. Accessed at http://honearchive.org/bibliographical/annotated.html.

Jacobson, Ethel. "She Would Have Preferred to Gnaw on a Plump Child." *Chicago Tribune,* 10 November 1963, J3.

James, William. *Essays in Radical Empiricism.* London: Longmans, Green and Company, 1912.

———. *A Pluralistic Universe.* London: Longmans, Green and Company, 1909.

Johnson, Samuel. *Letters of Samuel Johnson,* vol. 2, 1775–1782. Edited by R. W. Chapman. London: Oxford University Press, 1952.

———. "The Life of Pope." In *Lives of the English Poets,* edited by G. B. Hill. Oxford: Clarendon Press, 1905.

Kieckhefer, Richard. *Magic in the Middle Ages.* Cambridge, UK: Canto Books, 2000.

King, Greg, and Penny Wilson. *The Fate of the Romanovs.* Hoboken, NJ: Wiley, 2005.

"Kissinger Returns." *Harvard Magazine,* 12 April 2012, http://harvardmagazine.com/2012/04/kissinger-returns.

Knapp, Caroline. *Pack of Two: The Intricate Bond Between People and Dogs.* New York: Dial Press, 1999.

Kogan, Judy. "From Russia, with Love." *Harvard Crimson,* 25 February 1976, www.thecrimson.com/article/1976/2/25/from-russia-with-love-pby-the/.

Köhler, Joachim. *Richard Wagner: The Last of the Titans.* New Haven, CT: Yale University Press, 2004.

Kunze, Michael. *Highroad to the Stake: A Tale of Witchcraft.* Translated by William E. Yuill. Chicago: University of Chicago Press, 1987.

Kuzniar, Alice A. *Melancholia's Dog: Reflections on Our Animal Kinship*. Chicago: University of Chicago Press, 2006.

Lacan, Jacques. "The Seminar of Jacques Lacan, Book IX: Identification, 1961–62." Translated by Cormac Gallagher. Accessed at www.scribd.com/doc/59939681/Lacan-Seminar-IX.

Laërtius, Diogenes. *The Lives and Opinions of Eminent Philosophers*, vol. 6, *The Cynics*. Fordham University Ancient History Sourcebook, www.fordham.edu/halsall/ancient/diogeneslaertius-book6-cynics.asp#Diogenes.

La Fontaine, Jean. "The Little Dog." In *La Fontaine's Tales: Imitated in English Verse*, vol. 1, 220–22. London: C. Chapple, 1814.

Landor, Walter Savage. *Letters and Other Unpublished Writings of Walter Savage Landor*. London: R. Bentley and Son, 1897.

Langdon, Gabrielle. *Medici Women: Portraits of Power, Love and Betrayal from the Court of Duke Cosimo I*. Toronto: University of Toronto Press, 2006.

Leslie, Charles Robert. *Autobiographical Recollections of the Late Charles Robert Leslie, R.A.* Boston: Ticknor and Fields, 1860.

Lewis, Naomi. "Animal Farming." *Observer*, 27 October 1963, 29.

Linton, E. Lynn. "Reminiscences of Walter Savage Landor." *Fraser's Magazine*, edited by James Anthony Froude and John Tulloch, July 1870, 113–20.

Lispector, Clarice. *A Breath of Life*. Translated by Johnny Lorenz. New York: New Directions, 2012.

———. *Selected Crônicas*. Translated by Giovanni Pontiero. New York: New Directions, 1996.

Lodge, Meghan. "I Love My Dog More Than I Love My Husband." *Dogster*, 17 January 2013, www.dogster.com/lifestyle/i-love-my-dog-more-than-husband.

"Lord Castlereagh." *Blackwood's Edinburgh Magazine*, April–September 1817.

Lustig, T. J. "James, Arnold, 'Culture,' and 'Modernity'; or, A Tale of Two Dachshunds." *Cambridge Quarterly* 37, no. 1 (March 2008), 164–93.

Maeterlinck, Maurice. *Our Friend the Dog*. Translated by Alexander Teixeira de Mattos. Illustrated by Cecil Alden. New York: Dodd, Mead, 1913.

Malcolm, Janet. *Two Lives: Gertrude and Alice*. New Haven, CT: Yale University Press, 2007.

Mann, Thomas. *Bashan and I*. Translated by Herman George Scheffauer. New York: Pine Street Books, 2003.

Manso, Peter. *Mailer: His Life And Times.* New York: Simon and Schuster, 2008.

Marsden, Jonathan. *Victoria & Albert: Art & Love.* London: Queen's Gallery, 2010.

Martial [Marcus Valerius Martialis]. *Epigrams,* Book 1. Translated by Henry George Bohn. Bohn's Classical Library (1897), www.tertullian .org/fathers/martial_epigrams_book01.htm.

Maturin, Charles Robert. *Melmoth the Wanderer.* London: Penguin Classics, 2001.

Mayhew, Henry. *The London Underworld in the Victorian Period: Authentic First-Person Accounts by Beggars, Thieves and Prostitutes.* New York: Courier Dover Publications, 2012.

McCabe, James Dabney. *The Secrets of the Great City.* New York: National Publishing Company, 1868.

McGraw, Seamus. "$1.5 Million Paid for World's Most Expensive Dog." *Today,* 17 March 2011, www.today.com/id/42128943/ns/today-today_pets/t/ million-paid-worlds-most-expensive-dog/#.Uijei7z-GAM.

Meyrick, John. *House Dogs and Sporting Dogs: Their Varieties, Points, Management, Training, Breeding, Rearing and Diseases.* London: Jon Van Voorst, 1861.

Millan, Cesar, and Melissa Jo Peltier. *Cesar's Way: The Natural, Everyday Guide to Understanding and Correcting Common Dog Problems.* New York: Harmony, 2007.

Millgate, Michael. *Thomas Hardy: A Biography Revisited.* London: Oxford University Press, 2006.

Misztal, Mariusz. "Queen Victoria and Her Dogs." In *New Trends in English Teacher Education,* edited by A. Jesús Moya Guijarro and José Ignacio Albentosa Hernández, 385–95. Cuenca, Spain: Ediciones de la Universidad de Castilla–La Mancha, 2009.

Morris, Willie. *My Dog Skip.* New York: Vintage, 1996.

Moser, Benjamin. *Why This World: A Biography of Clarice Lispector.* London: Oxford University Press, 2009.

Nabokov, Vladimir. *Ada, or Ardor: A Family Chronicle.* New York: Vintage, 1990.

———. *Laughter in the Dark.* New York: New Directions, 2006.

————. *Lolita*. New York: Vintage, 1991.

————. *Speak, Memory*: *An Autobiography Revisited*. New York: Vintage, 1989.

Nicholas II and Alexandra. *A Lifelong Passion*: *Nicholas and Alexandra— Their Own Story*. Edited by Andrei Maylunas and Sergei Mironenko. Translated by Darya Galy. New York: Doubleday, 1997.

Nichols, William. *The Beloved Prince*: *A Memoir of His Royal Highness the Prince Consort*. Whitefish, MT: Kessinger Publishing, 2010.

Nixon, Richard. "Checkers Speech," 23 September 1952. Transcript, www.pbs.org/wgbh/americanexperience/features/primary-resources/ nixon-checkers/.

Nosowitz, Dan. "I Met the World's Smartest Dog." *Popular Science*, 26 August 2013, www.popsci.com/science/article/2013-07/i-met-worlds -smartest-dog.

Nye, Robert. "In the Night Forest." *Guardian*, 29 November 1963, 14.

Orel, Harold. *The Final Years of Thomas Hardy, 1912–1928*. Lawrence: University Press of Kansas, 1976.

Orpen, Adela E. "Royal Favorites." *Atalanta* 5 (1891), edited by L. T. Meade, 203–10.

Ouida [Marie Louise de la Ramée]. "Dogs and Their Affections." *North American Review* 153 (September 1891), 312–21.

Ovid [Publius Ovidius Naso]. *Metamorphoses*. Translated by Frank Justus Miller and G. P. Goold, 3rd ed. Cambridge, MA: Harvard University Press, 1977.

Palmer, Brian. "Would Your Dog Eat Your Dead Body?" *Slate*, 13 July 2011, www.slate.com/articles/news_and_politics/explainer/2011/07/would_ your_dog_eat_your_dead_body.html.

Petronius [Gaius or Titus Petronius Arbiter]. *Satyricon*. Translated by Michael Heseltine. Revised by E. H. Warmington. Cambridge, MA: Harvard University Press, revised edition 1969, reprinted with corrections 1987.

Petry, Alice Halle. "Caesar and the Artist in Willa Cather's 'Coming, Aphrodite!'" *Studies in Short Fiction* 23, no. 3 (Summer 1986), 311.

"Pets, and What They Cost." *Tait's Magazine*. Republished in *The Eclectic Magazine of Foreign Literature, Science and Art* 38 (May–August 1856), edited by W. H. Bidwell, 410–15.

Pirandello, Luigi. *Each in His Own Way and Two Other Plays*. Translated by Arthur Livingston. London: E. P. Dutton, 1923.

Plato. *The Republic*. Translated by Benjamin Jowett. Internet Classics Archive 2000, http://classics.mit.edu/Plato/republic.mb.txt.

Plutarch. *The Life of Alexander*, vol. 7. Translated by Bernadotte Perrin. Cambridge, MA: Harvard University Press, 1919.

Pope, Alexander. *The Rape of the Lock and Other Major Writings*. London: Penguin, 2011.

Porée, Adolphe-André. *Histoire de l'abbaye du Bec*. Paris: Charles Hérissay, 1904.

Pronek, Neal. *How to Raise and Train a French Bulldog*. Neptune, NJ: T.F.H. Publications, 1965.

Rabelais, François. *Gargantua and Pantagruel*, Book 1. Project Gutenberg, (EBook #8166), www.gutenberg.org/files/8166/8166.txt.

Rayfield, Donald. *Anton Chekhov: A Life*. Chicago: Northwestern University Press, 1988.

Richardson, Angelique. "The Many-Sided Thomas Hardy." *Times Literary Supplement*, 10 July 2013, www.the-tls.co.uk/tls/public/article1285627.ece.

Roberts, Jane. *Royal Artists, from Mary Queen of Scots to the Present Day*. Los Angeles: Grafton, 1987.

Roosevelt, Franklin Delano. Campaign dinner address ("Fala Speech"), 23 September 1944. Transcript, www.wyzant.com/help/history/hpol/fdr/fala.

Ross, Estelle. *The Book of Noble Dogs*. New York: The Century Co., 1922.

Rush, Richard. *Memoranda of a Residence at the Court of London*. Philadelphia: Carey, Lea and Blanchard, 1833. Accessed at www.archive.org/stream/courtoflondonoorushrich/courtoflondonoorushrich_djvu.txt.

Russell, Jeffrey Burton. *Witchcraft in the Middle Ages*. Ithaca, NY, and London: Cornell University Press, 1972.

Sacks, Jeffrey, Leslie Sinclair, and Julie Gilchrist. "Breeds of Dogs Involved in Fatal Human Attacks in the United States Between 1979 and 1998." *Journal of the American Veterinary Medical Association* 217, no. 6 (15 September 2000), 837–40.

Schlocker, Georges. "Literary Harvest in France." *Books Abroad* 37, no. 2 (Spring 1963), 148–51.

Schopenhauer, Arthur. "Ideas Concerning the Intellect Generally and in All Respects." In *Parerga and Paralipomena*, vol. 2. Translated by E. F. J. Payne. Oxford University Press, 1974.

———. *Studies in Pessimism*. Edited and translated by Thomas Bailey Saunders. London: Swan Sonnenschein, 1908.

Scott, Sir Walter. *Guy Mannering*. Edited by P. D. Garside. London: Penguin, 2003.

———. *Memoirs of Sir Walter Scott*, vol. 8. Edited by J. G. Lockhart. Edinburgh: Black, 1869.

Serpell, James. *In the Company of Animals: A Study of Human-Animal Relationships*. Cambridge: Cambridge University Press, 1996.

Shakespeare, William. *King Lear.* London: Arden, 1997.

———. *Macbeth*. London: Methuen, 1985.

Shaw, Vero. *The Illustrated Book of the Dog*. London, Paris, and Melbourne: Cassell and Company, 1890.

Slote, Bernice. Appendix to *Uncle Valentine and Other Stories: Willa Cather's Uncollected Short Fiction, 1915–1929*, edited by Slote, 177–81. Lincoln: University of Nebraska Press, 1973.

Smart, Christopher. *The Poems of the Late Christopher Smart*, vol. 2. London: Smart and Cowslade, 1791.

Souhami, Diana. *Gertrude and Alice*. New York: Pandora, 1991.

Stein, Gertrude. *The Autobiography of Alice B. Toklas*. London: Grey Arrow Edition, 1960.

Stewart, Jules. *Albert: A Life*. London: I. B. Tauris, 2011.

Stock, Jon. "A Dog's Life." *Independent*, 25 June 1995, www.independent.co.uk/arts-entertainment/a-dogs-life-1588260.html.

Stuart, Andrea. *The Rose of Martinique: A Life of Napoleon's Josephine*. New York: Grove, 2005.

Tannen, Deborah. "Talking the Dog: Framing Pets as Interactional Resources in Family Discourse. Research on Language and Social Interaction." In *Family Talk: Discourse and Identity in Four American Families*, 46–69. Edited by Deborah Tannen, Shari Kendall, and Cynthia Gordon. New York: Oxford University Press, 2007.

Tibol, Raquel. *Frida Kahlo*. Albuquerque: University of New Mexico Press, 1993.

Tomalin, Claire. *Thomas Hardy.* New York: Penguin, 2007.

Trubshaw, Robert Nigel, ed. *Explore Phantom Black Dogs.* Avebury, UK: Heart of Albion Press, 2005.

Turner, Pamela S., and Yan Nascimbene. *Hachikō: The True Story of a Loyal Dog.* Boston: HMH Books for Young Readers, 2009.

Updike, John. "Nabokov's Look Back: A National Loss." *Life,* 13 January 1967, 12–15.

Vane, Charles. Preface to *Memoirs and Correspondence of Viscount Castlereagh, Second Marquess of Londonderry,* vol. 1, *The Irish Rebellion.* London: Henry Colburn, 1848.

Victoria. *The Letters of Queen Victoria: A Selection from Her Majesty's Correspondence Between the Years 1837 and 1861.* Edited by Arthur Christopher Benson, M.A., and Viscount Esher, G.C.V.O., K.C.B. Whitefish, MT: Kessinger Publishing, 2006.

Villavicencio, Monica. "A History of Dogfighting." NPR, 19 July 2007, www.npr.org/templates/story/story.php?storyId=12108421.

Virgil. *Eclogues. Georgics. Aeneid: Books 1–6.* Translated by H. Rushton Fairclough. Cambridge, MA: Harvard University Press, 1919.

Volta, Ornella. "Give a Dog a Bone—Some Investigations into Erik Satie." Translated by Todd Niquette. *Revue Internationale de la Musique Française* 8, no. 23 (1987).

von Arnim, Elizabeth. *All the Dogs of My Life.* London: Random House, 1995.

Wagner, Richard. *My Life,* vol. 1. Project Gutenberg (EBook #5197), www.gutenberg.org/cache/epub/5197/pg5197.html.

Wallace, William, and John Parker Anderson. *Life of Arthur Schopenhauer.* London: W. Scott, 1890.

Watts, Isaac. "Let Dogs Delight to Bark and Bite." In *The World's Best Poetry,* vol. 1. Edited by Bliss Carman et al. Philadelphia: John D. Morris, 1904.

Wharton, Edith. *The House of Mirth.* New York: Charles Scribner's Sons, 1922.

———. "Kerfol." *Scribner's Magazine* 59 (1916), 329–82.

White, David Gordon. *Myths of the Dog-Man.* Chicago: University of Chicago Press, 1991.

Wicks, Robert. "Arthur Schopenhauer." In *The Stanford Encyclopedia of Philosophy* (Winter 2011 Edition), edited by Edward N. Zalta, http://plato.stanford.edu/archives/win2011/entries/schopenhauer/.

Wilby, Emma. *Cunning Folk and Familiar Spirits: Shamanistic Visionary Traditions in Early Modern British Witchcraft and Magic.* Brighton, UK: Sussex Academic Press, 2005.

William, Sister Mary. "Review of *Behind the Bathtub*, by Colette Audry." *Best Sellers*, 15 November 1963, 297.

Wittgenstein, Ludwig. *Philosophical Investigations.* Translated by G. E. M. Anscombe, P. M. S. Hacker, and Joachim Schulte. London: John Wiley and Sons, 2010.

Wolff, Cynthia Griffin. Introduction to *Coming, Aphrodite! and Other Stories*, edited by Margaret Anne O'Connor. New York: Penguin, 1999.

Woodward, Daisy. "Lucian Freud's Whippet." *AnOther*, 20 July 2012, www.anothermag.com/current/view/2070/Lucian_Freuds_Whippet.

Woolf, Virginia. *The Diaries of Virginia Woolf*, vols. 1–5. Edited by Anne Olivier Bell. New York: Harcourt, Brace, Jovanovich, 1977–84.

———. *Flush: A Biography.* Boston: Mariner Books, 1976.

———. *The Letters of Virginia Woolf*, vols. 1–6. Edited by Nigel Nicholson and Joanne Trautmann. New York: Harcourt, Brace, Jovanovich, 1975–80.

———. *Mrs. Dalloway.* Oxford University Press, 2000.

Wyndham, Francis. "Review of *Douchka*, by Colette Audry." *New Statesman*, 1 July 1963, 708.

Wynn, M. B. *The History of the Mastiff Breed, Gathered from Sculpture, Pottery, Carving, Paintings, and Engravings; also from Various Authors, with Remarks on the Same.* Melton Mowbray, UK: William Loxley, 1886.

Youatt, William. *The Dog.* Edited by Elisha Jarrett Lewis. London: Lea and Blanchard, 1845.

———. *The Obligation and Extent of Humanity to Brutes: Principally Considered with Reference to the Domesticated Animals.* London: Longman, Orme, Brown, Green and Longman, 1839.

Zvereva, Nina. *Avgusteyshie Sestry Miloserdiya.* Moscow: Veche, 2006.

About the Author

MIKITA BROTTMAN, PH.D., is an Oxford-educated scholar, critic, and psychoanalyst. Her articles have been published in the *American Journal of Psychoanalysis, Psychoanalytic Review, American Imago,* and elsewhere. Her previous books include *Hyena* and *Thirteen Girls.* She is Professor of humanities at the Maryland Institute College of Art in Baltimore.